Strands

By the same author

Tattoos for Mother's Day
Hard Water
Tilt

Jean Sprackland

Strands

A Year of Discoveries on the Beach

JONATHAN CAPE
LONDON

Published by Jonathan Cape 2012

2 4 6 8 10 9 7 5 3 1

First published in Great Britain in 2012 by
Jonathan Cape
Random House, 20 Vauxhall Bridge Road,
London SW1V 2SA

www.randomhouse.co.uk

Addresses for companies within The Random House Group Limited can be found at:
www.randomhouse.co.uk/offices.htm

The Random House Group Limited Reg. No. 954009

A CIP catalogue record for this book is available from the British Library

ISBN 9780224087452

The Random House Group Limited supports The Forest Stewardship Council
(FSC®), the leading international forest certification organisation. Our books
carrying the FSC label are printed on FSC® certified paper. FSC is the only
forest certification scheme endorsed by the leading environmental organisations,
including Greenpeace. Our paper procurement policy can be found at:
www.randomhouse.co.uk/environment

Typeset in Quadraat by Palimpsest Book Production Limited,
Falkirk, Stirlingshire

Printed and bound in Great Britain by
the MPG Books Group, Bodmin, Cornwall

For Nigel

A breeze was blowing, and I could smell salt, seaweed, and sun-bleached shore. I knew, once again, that I'd found home.

— Sue Hubbell, Waiting for Aphrodite

Contents

Preface   xi

SPRING
The *Star of Hope*   3
Mermaid's Purse   23
Prozac   30
Gooseberries and Jelly   35
The Underworld   49
Black Gold   54

SUMMER
Denatured   71
Poor Man's Asparagus   85
Aphrodite   89
The Albatross and the Toothbrush   103
Swarm   119
Old Seafarer   122

AUTUMN
Drifters   129
Sea Potato   140
Come in, Number 189   143
Stella Maris   147

Squirt   156
Queen's Cup   161

WINTER
Deep Freeze   179
SOS to the World   183
First-footing   197
Rough Lords   202
White Horses   215
Time Travel   218

Postscript   236
Acknowledgements   238

Preface

I've been walking on this beach for twenty years. In that time, our relationship has grown complex and intimate. It has become, as places can, an inner as well as an outer landscape, one I carry around in my head and explore in my imagination even when I'm far from home. It's been the setting for dreams, and it has seeped into my poems, as the backdrop for apocalyptic storms and miraculous walking on water, for journeys along the seabed and hauntings in the sand dunes. As soon as I sit down to write, this place is there, waiting. The version I carry in my head is endlessly flexible, but of course the external place does not obey me at all. It remains stubbornly unknowable.

I never tire of the coast because it's never the same twice. The tides and the weather change its physical shape, and they bring different things to look at. There's always something new. Perhaps if I had settled in a landscape of fells and lakes unchanged for thousands of years, my overriding sense of my environment – my apprehension of the natural world – would be of something solid, immutable, reliable, fixed. But this stretch of coast has an entirely different spirit. It's all about change, shift, ambiguity. It reinvents itself. It has a talent for concealment and revelation. Things turn up here; things go missing.

Over the years, I've made finds on my walks which sparked my curiosity. But I rarely followed up. I was busy; I forgot; something else grabbed my attention and the moment was lost. Now, though, life is about to change, and my time with this beach will soon be over. I'm getting married again, and a move to London is on the cards. In this final year before I leave, I want to honour this place by looking more closely and recording what I see. I want to take the time to search for answers to my questions, and to follow wherever that search might lead.

This will be a kind of travel book; but the travel will all be on foot, and will revisit again and again the same small area, covering just a few miles of the north-west coast of England: a stretch of shore book-ended by Southport Pier to the north and Formby Point to the south. In this endeavour I'm inspired and intimidated in equal measure by some very illustrious examples. Gilbert White spent a lifetime observing and studying the wildlife in one single parish – Selborne in Hampshire – which covered just a few thousand acres; and the French classic *A Journey Around My Room* is an account of forty-two days the author, Xavier de Maistre, spends exploring his own small living space while under house arrest. Both books cover very little ground, geographically speaking; but both writers approach their subject with what Alain de Botton calls the 'travelling mindset', which is characterised by curiosity and receptivity. 'If only we could apply a travelling mindset to our own locales,' he says, 'we might find these places becoming no less interesting than the high mountain passes and jungles of South America'.

Of all the British coastline, this is hardly the prettiest or the most unspoilt: its sands are not the most golden, and there are no rock-pools or hidden coves. Neither is it the most dramatic: no pounding surf, no rugged cliffs. Low tide can take the sea nearly two miles from shore. Stand on the beach at Ainsdale, on any reasonably bright day, and you can see the offshore wind farm in the Mersey estuary. Turn and face the other way, and there are the familiar Blackpool landmarks: the tower, the rollercoaster. But in really clear weather the bigger picture is visible: the southern fells, the Clwyddian hills, the pale but unmistakable shape of Snowdon. This is a place of big skies and lonely distances, a shifting palette of greys and blues; a wild, edge-of-the-world place.

I'm embarking on this year of discovery not as a naturalist, a historian or a geologist, but as an ordinary walker. I'm setting out, armed with curiosity rather than expertise, to pay a different kind of attention to what I see. I hope to cut through the blur of familiarity, and explore this place as if for the first time. Some of my finds may be real surprises, and others more predictable; but I shall pick them up and hold them to the light, regardless.

Spring

The Star of Hope

On a good day, cycling along the sea's edge is all exhilaration – sun flashing on moving surfaces, waves licking your wheels, gangs of gulls and sandpipers scattering into the sky just ahead.

But the hard road of sand I try to follow often gives way abruptly to soft, sludgy patches. The wheels bog down quickly and I have to jump off quick and haul us back on track. Then there's the likelihood of getting stranded the wrong side of a channel of water stretching for miles along the beach, separating me from the safety of land. I take a good run at it, hoping to power through on speed and sheer chutzpah. It can work, but sometimes the water is deeper than it looks. More than once I've misjudged so badly, the bike has come to a graceless halt right in the middle of a fast-flowing river, kept its balance a second, then fallen flat on its side, depositing me in two feet of water.

On this dazzling April afternoon, I've cycled through the pine-woods to Fisherman's Path, then dragged my bike through the soft sand between dunes, emerging to find the tide low and the beach remade in new shapes. There's a long sandbank or bar, perhaps twelve feet wide and three feet high. I've seen something like this before, and I've walked along it as if walking a city wall or castle rampart, with a moat of beached seawater below.

This is what keeps me coming back. This shore is wild and

changeable, obeying no rules, refusing to be subjugated. It's a shape-shifting place, in league with the wind and the moon and other forces of unimaginable power and energy. I arrive, and the terrain has been reinvented: sometimes ridged, sometimes dead flat, sometimes gouged by deep runnels of swift water which the sea is sucking back. There are few fixed landmarks here: you have to stand with your back to the sea and find some onshore feature by which to orientate yourself, but even this is unreliable. That high sand dune that looks exactly like a camel's hump may not be quite the same shape next time you look. There's a caginess, a reluctance to be pinned down or identified, a sense in which this place could give you the slip.

But today is exceptional. There has been a great upheaval. The landscape is transfigured, and there are three wrecked ships lying on the surface.

From this distance they don't look real. The beach is like a stage, set with elaborately made props. It's as if they've been towed onto the sand and installed here overnight, perhaps in some large-scale public art project to rival Antony Gormley's iron men down the coast at Crosby.

*

Children know all about the magic of hidden realities beneath the surfaces of things. They know it intimately, from experience, and it's reinforced by the stories that surround them in books and films. Every child who ever read C.S. Lewis has climbed into at least one wardrobe and felt around for the secret doorway into the other world beyond. The streets and paths and woods near my childhood home were familiar to me in an intensely physical way which no place has achieved since, but underneath their obvious surfaces there were other, secret dimensions I explored alone. One of these was behind a grille on the pavement outside the grocer's shop: if I knelt and peered through the grille in just the right way, I was sure I could make out a gloomy underground room, a secret bunker, full of broken dolls. They lay in a jumbled heap, their vacant faces upturned to the street and the daylight above. As an adult I don't know how to make any sense of this vivid, exciting memory, but the scene on the beach today reminds me of the vital feeling I experienced as I knelt on the tarmac with my six-year-old cheek pressed to the cold metal: *There is more to things.*

I feel that same thrill today, standing here and looking at these three broken old ships. I know that these sands shift, playing an endless game of hide-and-seek. Even so, it's hard to believe that objects as large and interesting as these have been here all along, shallowly buried, and I've cycled over them, oblivious.

I haul my bike over the high sandbar and cycle to the first wreck, far out near the sea's edge. It's a wooden vessel, very well exposed, with a sturdy post which might have been a mast, and curved wooden spurs like the ribcage of some extinct beast, picked clean of its flesh by the sea.

I cycle on to the second, an old favourite. It's the *Star of Hope*, a wreck I've visited several times on occasions when the sand has yielded it. It's lying forlornly in muddy water, heavily barnacled, black and rotting in places.

The third, further north and closer to shore, is huge and listing, spilling its cargo of wet sand. It's a more solid sort of craft, and I've never seen it before. The deck is missing, but there is a framework of bent spars with iron knees which must once have held it together and now give some idea of its size. There's a contraption which looks like a windlass for winching freight on board or for raising the anchor.

When I say that this place is changeable, this is one of the things I mean. The tides and currents conspire to move and reshape the sand, and in the intertidal zone a skeletal old ship emerges. The rest of the time it's buried and invisible; people and dogs walk and run over it with no idea that it's there. Then, without warning, it rises to the surface. It takes the air for a few weeks, before subsiding into the sand again.

I first experienced this on a spring day five years ago, when a friend called in a state of high excitement and read out an article from the local paper, under the marvellous headline *Boat sunk in storm rises again*. According to the report, it was the first time in seven years that the *Star of Hope* had come so completely to the surface. Once wrecked, she was claimed by the sands, which stowed her away underground, working her to the surface only occasionally. No one knew how long she would remain visible, so my friend and I arranged to meet on the beach, along with our teenage sons, and catch a glimpse of this phenomenon while it lasted.

Until the sands shift and reveal it, the *Star of Hope* is sealed in its sandy tomb. From time to time there are tantalising clues: sometimes the place is indicated by a group of wooden stumps sticking up out of the sand at low tide, like grave markers, squatted by a couple of cormorants with hunched black shoulders and reptilian necks. I'd walked out to these mysterious stumps dozens of times, kicked them experimentally and found them solid, speculated about what might lie below the surface.

We walk over the remains of the past all the time: what's left of our ancestors, their buildings and artefacts, traces of their food and clothing, tools and toys – the 'physical culture', as archaeologists call it. But mostly we're so preoccupied with the here-and-now that we don't think about what lies berneath the surface we're standing on. Here on the beach, walking around and between the rough, worn wooden spars, I've had moments of acute awareness of that other, buried reality, just a few metres below but completely inaccessible. I'd suddenly have the same vertiginous feeling I experience when I sail in a boat on deep water, and think about the shadowy space beneath: there is another dimension, and it's full of mysteries. It's dizzying, that realisation that we spend our lives moving precariously on the outer skin of the planet, and that same skin contains all the stuff of history.

I'm never more conscious of this than when I'm walking on the beach. The sand I print with my boots had a former life as rocks and mountains, and as I walk I crush the shells of dead sea-creatures underfoot.

*

Ships are not the only things lost under these sands. A few months ago I was walking back along the beach with a friend when he stopped and pointed out a ridge of yellow painted metal curving up just proud of the surface sand.

I often find scrap metal here: sometimes recognisably a pipe, the broken blade of a saw, part of a gate, but more often miscellaneous bits and pieces with very little about them to hint at their provenance.

The unusual shape and colour of this object had us intrigued. We got hold of it and tried to pull it out, but it was stuck fast. We dug away with our bare hands and with a stick, but the sand was wet and solid, and the thing was obviously part of some much bigger thing, and it wouldn't budge – not even a millimetre. It looked vaguely familiar, but I couldn't make sense of it. My friend stopped digging and took a step back to reassess it.

It's a car, he said.

Suddenly I recognised that metal clue for what it was: the curved frame of a car door. We paced out the potential shape of the whole vehicle. A few other knobs and knuckles on or just below the surface gave it away. Even so, it wasn't easy to map its size and shape; it was a bit like reaching up into a high cupboard and feeling around, trying to visualise what's in there. Then beginning to realise that it's not a tin of beans or a jar of peanut butter but something incongruous like a cat or a set of false teeth.

I'd heard that there were shipwrecks buried here, but I hadn't thought of a car. What else might be under there? A crashed helicopter, an abandoned train?

The sea that calm day seemed so gentle, the tide ambling quietly up the beach, that it was difficult to imagine it powerful enough to pick up large heavy objects like cars and ships, or to move vast quantities of sand, enough to cover them. But when the westerly gales howl in, this place has quite a different character. Countless times I've seen the shore hewn and hammered, scattered with whole tree trunks, steel pipes, oil drums, concrete fence posts, dead sheep. The very topography of the beach is modified by these batterings, so much so that I have walked an altered landscape and thought of the phrase 'sea change', first spoken by Ariel in The Tempest:

Full fathom five thy father lies;
Of his bones are coral made:
Those are pearls that were his eyes:
Nothing of him that doth fade,
But doth suffer a sea-change
Into something rich and strange.

The phrase is generally used to mean a profound or fundamental change, something like a U-turn, but Brewer's *Dictionary
of Phrase and Fable* has it as 'an apparently magical change, as
though brought about by the sea'. A real and tangible version of
that magic is wrought here, over and over again. Because there
are no outcrops of rock on this coast, the forces of nature mould
and re-mould the shoreline. It's a dynamic landscape; the size
and variety of the debris, as well as the unpredictable shape of
the beach surface itself, are proof of the colossal kinetic power
of wind and waves.

Another time, I found a wrecked van. The police had got there
first; it was their tape flapping in the wind halfway down the
beach that attracted my attention. This is a place of choice for
twockers and joyriders; first they get the adrenalin rush of the
police chase through the streets and down the coast road, and
then they end up on these empty sands, where there's space to
drive, fast and reckless, practising handbrake turns and burning
out the gears.

It looked like the scene of a violent crime. The van was lying on
its roof, windows smashed, bodywork punched in with a blunt
instrument. It had been set on fire. Molten solder had dripped and
pooled on the sand like congealed blood, and steel fibres from
inside the tyres had been torn out in handfuls like hair. A few yards

away, a ripped-out seat lay on its side, the innards dragged and spread out under the wind and rain. Perhaps it would be gathered up, towed away and scrapped. If not, it too would start to work its way into the sand, disappearing little by little, until it was swallowed and lost.

*

Joyriders are not the only ones who want to claim the beach for the internal combustion engine. Motor vehicles have always had a complicated relationship with this beach. Until a few years ago, cars were permitted, free of charge, to be driven and parked on every part of the shore. When restrictions were introduced, limiting the area accessible by vehicle and introducing an entry charge during the summer months, there was uproar. These apparently modest changes to the bye-laws unleashed a torrent of righteous indignation. For months on end, the letters pages in the local press were full of furious protests from long-time residents claiming it was an infringement of their basic human rights. Years on, the rage has subsided into nostalgia: remember the good old days when you could drive right up to the edge of the sea?

Even beyond the barriers which mark the limits of the vehicular zone, the beach is not completely vehicle-free. There are people who are allowed access, either for work or other special purposes: a number of permits are issued to fishermen, for instance, so that they can tow their craft to the water's edge. Coast rangers and National Trust employees use Land Rovers to visit remote areas in the course of their management and conservation work. In summer, pale lifeguards strike somewhat unconvincing *Baywatch* poses on an amphibious vehicle, scanning the distant waves through binoculars. At the height of the season, the more populated areas can be like pedestrianised city-centre streets where shoppers have to negotiate taxis, delivery lorries and all manner of exempted vehicles. Nevertheless, the general driver, visiting the beach for leisure and pleasure, is confined to an area extending fifty metres or so either side of the beach entrance, and in the summer months must pay £3 for the privilege.

The restrictions were put in place as part of a new recognition that this stretch of coast was special and should be protected. I suppose there will always be a few people who don't understand why conservation matters, and rather more who don't want to have to pay even a modest price for it. But all that talk about rights seems to say something more general about our sense of ownership where beaches are concerned. There's a deeply felt conviction that they belong to the people, and that no one should be able to call that ownership into question, or to put limits or conditions on it. The beach is a place of freedom where we ought to be allowed to do whatever we like.

Joe Moran, writing in The Guardian, has described the beach as 'a frontier not only between water and solid ground, but also between the wild and the domestic', a place which condones 'a certain amount of low-level lawlessness, from nicking boulders for garden water features to scavenging for Nike trainers in the cargo ship containers that occasionally wash up'.

When a container ship was beached at Branscombe in Devon in January 2007, hundreds of people went to the beach and scavenged the goods they found washed up there, including brand-new BMW motorbikes, steering-wheel airbags, wine and electrical goods. At first the scavenging was tolerated, but as organised gangs moved in the police eventually sealed off the beach and threatened to invoke the Merchant Shipping Act of 1854 in an attempt to force people to return what they had taken.

There was high excitement in the local press, more accustomed to publishing headlines like *Pensioner trapped in bathroom* and *Potholes patched up as villagers stage protest meeting*. BBC pictures showed people incandescent with delight as they carried off packs of nappies and face cream. Their enthusiasm seemed all out of proportion to the value of the goods, but here was a heady combination of two very British pastimes: beachcombing, and getting something for nothing. It was as if the sea, as it carried the stuff from ship to shore, had washed it clean of ownership, leaving it freely available to be taken with a clear conscience.

The law, of course, would not agree: there is still an official Receiver of Wreck, to whom all finds of flotsam and jetsam should be surrendered (along with any dead whale, dolphin, porpoise or sturgeon found on our shores, as these have qualified, since the fourteenth century, as 'royal fish').

But even if this scavenging qualified as a crime, did it really matter? Surely at least it was more or less victimless? Not for the couple, in the process of emigrating to South Africa, who turned on *News at Ten* and watched as their personal possessions, clothing and family photographs were tipped out of a crate onto the beach by scavengers, quickly judged worthless, and left to be scattered by the wind and the tide.

I'm sure I wasn't alone, watching the Branscombe story unfold in the television news reports, in feeling a shudder of apprehension. These images prefigured those of the riots in English cities in the summer of 2011; people's faces, lit by torchlight and camera flash, were just like those of looters in times of civil war or natural disaster. This was 'passive wrecking' – not luring ships onto the rocks, *Jamaica Inn* style, but a willingness to profit when someone else's misfortune provides an opportunity.

★

There's an area of my local beach which has a reputation for illicit activity of other kinds. I refer to dogging, described by one website as 'a predominantly British activity that involves outdoor exhibitionism in car-parks, wooded areas and the like'. The internet appears to have changed the whole climate for transgressive sexual behaviour; even swingers publish mission statements these days. There is a 'dogging locations search tool', a section on dogging etiquette, and warnings about litter and inconsiderate parking. The sample pictures available for viewing by non-members are far from erotic to my eye, but that may be down to a combination of unflattering camerawork, uncomfortable locations and very British weather.

It's hard to judge how much truth there is to the local rumours. I rarely venture onto the beach at night, except in the spring, during the two or three weeks when natterjack toads are breeding in the sand dunes. The natterjack, with a distinctive yellow stripe down its back, has become the poster boy for the Sefton coast. The shallow slacks in the dunes make an ideal breeding ground, and half of the total extent of this habitat in England is here. The toads cling on to survival against formidable odds: slacks are dependent on groundwater levels and are highly sensitive to drainage and abstraction; if there is a long-term fall in water level the conservation of dune wetland is threatened. In the short term, a dry summer means a bad year. One of the coast rangers employed by the council told me that fifteen years ago, when he was first involved in natterjack conservation, he and his colleagues would go out with thirty-five-millimetre film canisters, collecting up the toadspawn from shallow puddles as

13

they dried out in the sun, and carrying it to larger and more suitable pools. Now, he said, the thinking has changed; they intervene less, they leave nature in charge and let the spawn take its chances. There's a larger picture, he reckons: bad years and good years balancing one species against another, maintaining diversity.

Natterjacks are known for the exceptionally loud nocturnal mating chorus of the males, a magical thing to experience. Last spring I took a friend and fellow poet with me. He'd always wanted to hear natterjacks, but he'd never made it until now, in spite of living his whole childhood just a few miles away on the outskirts of Liverpool. It was a clear, still April evening. We waited in a restaurant half a mile away until dusk fell, then set off along the beach towards a path into the dunes, a place where I've heard natterjacks before. As we walked, we passed the odd parked car here and there, and hurried past with our hoods up, reluctant to look in case we saw something we'd wish we hadn't.

We cut inland and stumbled through the dark into the rugged landscape of the dunes, stepping here and there on a discarded vodka bottle or the charred remains of a bonfire. It is, needless to say, forbidden in the bye-laws to light fires here, or to camp overnight. But the beach has an entirely different life at night, and there is something timeless about it. People – young people in particular – go there to escape supervision, rules and social conventions. When darkness falls they head for the wild frontier. And when they get there, their entertainments are the same as those of their distant ancestors: lighting fires, getting out of their heads, and shagging.

In the dunes, the natterjacks provided the nightlife. They were in good voice. We kept our torches switched off and our footsteps as soft as possible. The toads are hypersensitive to sound, or vibration, and they enforced an exclusion zone around each slack, by falling silent if we stepped across an invisible line and got too close. We edged our way to a wide flat place where we were surrounded by thousands of them, all calling together, throwing their crackling voices to the night sky like a clamour of electronic signals, a code that can carry for miles, and that reaches back into the prehistoric past. I don't know what was happening in those parked cars, but there was a riot of amphibian sex going on all around us, eternal and moonlit.

*

The Star of Hope was a three-masted barque, built entirely of timber, at the Stephen & Forbes yard in Peterhead. According to the dictionary, a barque is 'a sailing ship, typically with three masts, in which the foremast and mainmast are square-rigged and the mizzenmast is rigged fore and aft'.

I'm a bit of a landlubber, so I have to do a little more research to get a clearer understanding of this definition. It seems there are several distinct types of barque, but even after reading up I can't tell which kind the Star of Hope is – brigantine, perhaps, or barquentine, or better still the wonderfully named jackass-barque.

The barque was very popular in the mid-nineteenth century, mainly because it was easier to handle and required a smaller crew than a full-rigged ship. The Star of Hope was an ordinary working craft, plying a routine trade. She was launched in 1865 and wrecked on a treacherous sandbank graphically known as Mad Wharf, in a Force 10 gale in 1883. On the fateful night she was sailing under a German flag from Wilmington, North Carolina to Liverpool, with a cargo of raw cotton destined for the Lancashire textile mills, when she was caught in the storm. Captain Hamann and his eight-man crew abandoned ship and were rescued, but craft and cargo were a total loss.

It was just one catastrophe among many. This coast used to be notoriously difficult to navigate for vessels trying to make it to safe harbour at Liverpool. One of the sea charts held in the archives at Liverpool Maritime Museum shows just how narrow the shipping channels were, and how shallow – in places they are marked as just a few feet deep. Mariners had to steer their craft with great care to avoid being grounded or blown onto Mad Wharf, Great Burbo Bank, Askew Spit, Taylor's Bank, Spencer's Spit, Zebra Flats or Mockbeggar Wharf. By 1892, when this chart was made, lightships and lighthouses were deployed to make the passage less hazardous. An electric cable is marked, linking Formby Lifeboat House with Formby Lightship, which is described in a note as having 'revolving red light every 20 secs, red ball, steam foghorn 4 blasts in quick succession every minute'.

The chart also marks a sophisticated system of buoys, differently shaped and coloured to indicate their positions: black conical buoys on the starboard side, black can buoys on the port side; beacon buoys with perch and globe; spherical buoys with black and white horizontal stripes, with staff and diamond, with staff and triangle; pillar, bell and gas buoys marking Special Points in the channel.

Nevertheless, there were many accidents, including Britain's worst ever lifeboat disaster, in which twenty-seven lifeboatmen from Southport and St Anne's drowned trying to rescue the crew of a barque called the *Mexico* in a storm in 1886. The Fisherman's Rest pub, where the drowned bodies were temporarily laid after being hauled out of the sea, acts as a kind of shrine to this notorious event. There are sepia photographs of the lifeboatmen with their

16

improbable beards, oilskins and cork jackets. A framed poem documents the story. It begins:

Up goes the Lytham signal, St Annes has summoned hands,
Knee-deep in surf the lifeboat's launched abrest of Southport
    sands;
Half-deafened by the screaming wind, half-blinded by the
    rain,
Three crews await their Coxswains to face the hurricane.
The stakes are death or duty, no man has answered No!
Lives must be saved out yonder, on the good ship Mexico.

Given the frequency of disaster, and the huge commercial importance of the port of Liverpool, it's hardly surprising that ambitious work began around the turn of the century to dredge out the shipping channels and make navigation easier. These days, the estuary takes a quarter of all container shipping between the USA and the UK, and accidents are rare.

Liverpool was the source of another innovation of great maritime importance. The city was home to the pioneer of modern tide tables, those calculations used by shipping companies, fishermen and leisure boat users to predict the height of the tides in advance. A hundred years before this sea chart was drawn, a man called William Hutchinson became Dock Master, and took it upon himself to measure and record the heights and times of high tides – a task he continued, day and night, and in all weathers, for thirty years. So meticulous were his methods that the data he collected are being used today in the study of climate change.

Hutchinson was a larger-than-life character, at various times a shipowner, boatbuilder, commercial trader, local politician, inventor, author and philanthropist. He sailed all over the world, first as a cook's cabin boy and later as a privateer in the Seven Years War. He was said to be aggressive in battle but strictly religious in character, and he tolerated no swearing or blaspheming, which must have made life difficult on board ship. He observed a day of prayer every year, on the anniversary of an occasion when he and his starving crew were delivered after being shipwrecked on a barren coast. They had just drawn lots to decide which of them should be put to death in order to feed the others. Hutchinson had lost the draw, and had submitted to the inevitable, but was saved at the eleventh hour by the arrival of a rescue ship.

As well as his groundbreaking work on tides, Hutchinson invented special reflecting mirrors and oil-burning lamps for lighthouses; established the world's first lifeboat station; invented a new type of ship's rudder, and a priming mechanism for guns; cut a channel into the Mersey estuary near New Brighton; founded a charity for the benefit of seamen's widows and children; and worked with a doctor at Liverpool Infirmary to pioneer early methods of artificial respiration.

Yet of all his dizzying array of achievements, the detail I love best is his invention of a special method of making tea on board ship. The apparent domesticity of this innovation is what touched me first, but for Hutchinson tea was not a cosy ritual but a lifesaver. In his *Treatise on Naval Architecture*, published in 1794, there is a chapter entitled 'On the Sea Scurvy', in which he gives an account of his experience of this appalling and intractable disease. He describes in horrifying detail its progress, starting with pain in the breast, blackness in the armpits and swollen limbs, and progressing inexorably until all he can do is lie in his hammock, pining away 'with my teeth all loose, and my upper and lower gums swelled and clotted together like a jelly, and they bled to that degree, that I was obliged to lie with my mouth hanging over the side of my hammock, to let the blood run out, and to keep it from clotting so as to choak me'.

Unlike many of his shipmates, however, Hutchinson recovered. He immediately set about researching the causes of scurvy, and working to develop preventive measures, one of which was the drinking of tea:

This attack of this destructive disease, appeared evidently to proceed from eating too much salt meat, and a short allowance of water &c. I therefore formed a resolution, as much as possible, to endeavour to avoid it in future, by eating no more salt meat than just to give a small relish to my bread . . . and at all possible opportunities, to get a little tea made for breakfast . . . And to let lovers of tea know, I have had it made in a very nice manner on board a ship, where there were no tea utensils; I will inform them that it was by putting the tea into a quart bottle, filled with fresh water, corked up and boiled in the ship's kettle, along with the salt beef. When it came out, it was as fine drawn tea as ever I saw. It may afterwards be sweetened at pleasure, in the same bottle; and with something to eat, may be given to sick people in bed, to bite and sup, in stormy weather, when teapots, cups and saucers, and other means cannot be used with safety.

*

Standing looking out to sea can be a bit like standing at an altar. You wait in silence for some sort of benediction. If prayer could have a physical destination, this would be it: not the sky but the

19

sea. Matthew Arnold may have heard the 'melancholy, long, with-drawing roar' of faith in the sound of the waves on Dover Beach, but I think there is still one of the ancient gods out there: a proper, old-fashioned, ineffable god, in the deep space we have still not quite fathomed. Still a maker, still a wrecker.

Hundreds of vessels have come to grief on this coast, forced onto hidden sandbanks, capsized, broken. Many of them are recorded in a chronicle known as the 'Lawson Booth List', and a local expert, Martyn Griffiths, has made a lifetime's study of them.

For thirty years, Martyn has been researching and identifying some of the wrecks, including the three I saw today. The third one, with the iron knees, is a nineteenth-century vessel called the *Atlantic*, which came to grief on its way from Liverpool to North America with a cargo of salt, and only began to emerge from its sandy grave this year. The first, with its tall mast, is still a mystery: it's probably of a similar vintage, but no one knows its name or anything about it. And there are others: the *Chrysopolis*, the *Zelandia*, the *Ionic Star*, the *Nerus*. Some can be reached only at an exceptionally low tide, and the shore can be a dangerous place at these times. The tide races in over the flat sands, at speeds of up to three metres per second. Distances are difficult to judge, and it's easy to become disorientated, especially if the weather changes. Anyone tempted to take risks here would do well to remember we're less than fifty miles from Morecambe Bay, where twenty-one Chinese cockle-pickers were drowned by an incoming tide in 2004.

The odd ship still ran into trouble on this coast, well into the twentieth century. The *Charles Livingston* was caught in a storm off Ainsdale in 1939, with the loss of twenty-two men. Six others survived by lashing themselves to the mast for ten hours until they could be rescued. And in 1950, a wooden schooner called the *Happy*

Harry had a run of particularly bad fortune: first it was grounded on Taylor's Bank and rescued by lifeboat, then it was refloated and sailed to Southport, where it promptly crashed into the pier. It met an undignified and distinctly unhappy end, dismantled and burnt in a huge bonfire on the beach.

Other wrecks have had bizarre afterlives. A steam trawler called the *Endymion* was used for target practice during the Second World War. The tall mast of the *Pegu*, a passenger ship wrecked en route from Glasgow to Rangoon, was for years a familiar local sight poking up above the surface in the shallows, acting as a perch for seabirds, before a tug collided with it in 1987 and demolished it. An unknown wreck rested for years in the sandhills, before being dragged back down the beach, floated out to sea and blown up by the Royal Engineers.

Visiting one of the local wrecks is always exciting. But today is spectacular, a one-off. I don't expect I'll ever have another chance to see three of these hoary old survivors in one go: ramshackle, barnacled but still standing. It's as surprising and memorable as that raw February afternoon when I saw the *Star of Hope* for the very first time. There was an abrasive wind, and the sea was flattened to a sullen grey line on the horizon. But the wreck was an astonishing sight, sitting on the sand in a shallow pool of water like an overgrown toy boat in a puddle. The boys were sloshing through in no time, scrambling onto the sloping deck like marauding pirates. My friend and I followed a bit more cautiously, and slithered across from port to starboard. We touched the handrail, peered at the four or five holds, the hole where the mainmast used to be. There was a stink of seaweed and rot, but the boat had spent much of its time vacuum-packed in sand, which had preserved the timbers remarkably well.

And there we were, the four of us: standing aboard a Victorian cargo vessel, marooned half a mile from the sea's edge. It broke a recurring dream, in which I'm on a ship which is landlocked and going nowhere. It's one from that genre of frustrated, impotent dreams in which nothing works as it should: a clock without hands, train doors which will not open, a knife and fork which crumple like paper.

The *Star of Hope* has her own, very curious afterlife. She's been sinking and rising, sinking and rising for over a century, in a ghostly reprise of that first calamity. The trauma is re-enacted time and again, in sand instead of water, and without her cargo or her crew.

Mermaid's Purse

The origin of the word 'purse' is in the Greek word byrsa, meaning hide or leather; and the little black parcels which are often to be found in the strandline look very much like black leather, scuffed or glossy.

These are mermaid's purses, the egg cases of dogfish, ray and skate. A closer look reveals that they're not all the same. Some are almost square, with a long, curved spike at each corner: these belong to ray and skate. Others are more oblong in shape, with fine, curled tendrils: these are the purses of dogfish. It's not at all unusual to find both types together on this beach, and sometimes there are scores of them, cast like small, cryptic gifts all along the strandline.

The folk name 'mermaid's purse' marks these out both as enchanted objects and as things of childhood. My own children used to love to collect them, and when she was six my daughter – always fascinated by the names of things – took one home and insisted on using it as a purse for small coins. But enchantment has its darker side – an alternative name is 'devil's purse', and on a February afternoon, with a sky turning navy-blue and a cold wind cutting in from the sea, it's not difficult to recast them as sinister objects. Either way, they have something of the fairytale about them, these jet-black parcels, sealed and mysterious, with their frills and curlicues.

The word 'dogfish' is used to describe a number of species of small shark which inhabit Northern European oceans. There are over four hundred species of shark in the world, ranging in size from fifteen centimetres to over six metres long. Many species give birth to live young, but of those which lay eggs, each has its unique purse or egg capsule, some fantastic in appearance. Most have either the points or curled strings I've observed on the ones I've found, but the capsule of the horn shark is corkscrew-shaped; the female screws or wedges it into a suitable crevice in the rock, and the spiral flanges keep it in place and out of the way of predators for the six to ten months it takes to hatch. In photographs, it's an alien-looking object, more like a small but vital part a mechanic might pull out of the back of a washing machine and shake his head over: 'Sorry love, your capsule's gone.'

*

Here on my local beach, I've only ever found two varieties – square and oblong – though these basic shapes actually contain a much richer variety. The Shark Trust provides an identification guide, explaining how to tell species apart by soaking and measuring, comparing the length of the distal and proximal horn, examining the edges for the presence or absence of a lateral keel. The guide covers the whole elasmobranch family, which includes skates and rays as well as sharks. I would very much like to find the common skate purse, which is about the size of an A5 notebook and covered with fibrous, pale golden 'bark' that peels off to reveal the shiny black surface underneath; but sadly, 'common skate' is now a misnomer – it appears on the Critically Endangered list and such a find is unlikely. I have more of a chance with the starry skate, whose purse has a special 'washboard' texture on one side to help it stick to the seabed. Or the cuckoo ray, very small and neat, with curved horns as long as the capsule itself.

The ones I find are mostly desiccated, blackened and hardened by their time out of water. Sometimes, though, you can find one freshly cast, and this will look quite different: soft in texture, olive brown in colour, translucent. The chances are it's empty, its occupant already swum free. But it's just possible, especially after stormy weather, to find one unhatched. Pick it up and hold it to the light, and you'll see the developing embryo inside. If you're very lucky, gestation will be complete, and you'll see a tiny baby skate or shark, almost ready to hatch. If you just happen to come along at the right time you can even assist at the birth. A fish-keeper called Jim Hall has been doing this for twenty years. He searches for freshly cast egg capsules of the lesser spotted dogfish on the beaches of South Wales, and takes them home to his aquarium. There he watches until the embryo is fully developed. At the right moment, he pinches the purse gently. If the hatchling struggles to break free, he snips the purse open. And there it is – a newborn sharklet, just a few centimetres long.

Other people's hobbies sometimes seem very peculiar, but I can see the element of challenge which could make this an absorbing activity. After early failures, Jim has learnt, by trial and error, how to raise fully grown specimens, and now he's hooked. Once the birth is successfully achieved, he has to tend the young fish carefully through its early days, while it lies about sluggishly on the gravel looking helpless and frail, until it's able to feed and fend for itself. He has documented the process on the website of the British Marine Life Study Society, and undoubtedly the main purpose is scientific research. But perhaps there's the additional satisfaction of saving life at the same time. After all, once the purse has been torn loose from its seaweed anchor and hurled onto the beach, exposure to the air will soon dry it out and kill the embryo. Collecting it and giving it a second chance in his aquarium is a small act of rescue.

Jim is not alone in his endeavours. In his book *Loch Creran: Notes from the West Highlands*, published in 1887, William Anderson Smith describes a similar experiment:

> Amid a mass of tangle attachments we found an egg of the Rough Hound, which we tossed into the 'live' bucket, thinking to keep it and bring it out. On re-examining it we found several young shells on the back of the egg, of an interesting character, and to obtain these we resolved to destroy the egg. Carefully cutting it open, the occupant was found to be fully formed, and placed in water it immediately uncoiled, and showed signs of life. The eyes were shut like a young puppy's, and it continued breathing through its mouth steadily, as if its gills were not sufficient to supply it . . . the little fish was the picture of the full-grown dog-fish, and the title of Roussette was even more strongly applicable to the prettily-spotted youngster than to the various full-grown specimens we have captured.

*

Saving individual fish is one thing. But it would require an unprecedented global rescue effort to stop whole species from catastrophe. Fishing on an industrial scale is emptying our oceans, and sharks are especially vulnerable, partly because of an insatiable appetite for their fins, and a scientifically discredited claim for the anti-cancer properties of their cartilage. They are also caught accidentally by long-lines, trawls and fixed tuna traps. Damage to marine environments in the course of fish farming and the clearing of mangrove forest is destroying shark nurseries, which live-bearing species depend on.

As if the odds were not stacked heavily enough against them, sharks have also suffered from human fears and superstitions, which have long made them an unpopular and deeply misunderstood cause. This has begun to change: the cliché of shark as killing machine – the monster with an insatiable lust for human blood – is gradually giving way to a more sympathetic and a more realistic image.

In Eiléan Ní Chuilleanáin's poem 'The Sun-Fish', basking sharks off the Irish coast are instead cryptic and mythical, hard to make out in the 'dappled light and foam':

> The salmon-nets flung wide, their drift of floats
> In a curve ending below the watcher's downward view
> From the high promontory. A fin a fluke
> And they are there, the huge sun-fish,
> Holding still, stencilled in the shallows.

Elusive as it can be, the sunfish is just one of numerous vulnerable sharks. Of over 400 known species, 201 are endangered, according to the Red List published by the World Conservation Union. The red-tailed black shark is now extinct in the wild. The daggernose shark, Harrison's deepsea dogfish, the Pondicherry shark and the sawback angelshark are amongst those 'critically endangered'. Sliding towards oblivion are the dusky shark, the night shark, the bigeye thresher shark, the circle-blotch pygmy swell-shark, the saddled carpet shark, the bristled lantern shark, the happy eddie, the brown shyshark, the sickle fin weasel shark, the snaggletooth shark, the Papuan epaulette shark, the sharpnose sevengill shark, the pale ghost shark, the South China cookiecutter shark, the harlequin sharkminnow, the sweet william, the white-spotted gummy shark, the tawny nurse shark, the spotted wobbegong, the dwarf ornate wobbegong, the slender weasel shark, the pyjama shark, the crocodile shark, the spadenose shark, the frog shark, the little sleeper shark, the scalloped bonnethead.

Skates and rays are no strangers to the Red List either. We may lose for ever the longheaded eagle ray, the thorny skate, the Brazilian blind electric ray, the sleeper torpedo, the pointynose blue chimaera, the Madagascar numbfish, the ribbontailed stingray, the cinnamon skate, the barndoor skate, the slimeskate and the munchkin skate.

The naming of a species is a celebratory act – a moment of recognition, a valuing. No wonder these names are so elegant, funny and particular. Each marks a spike of optimism, when the world became, to human eyes, a richer and more contrapuntal place.

*

Names matter. Even if it wasn't an endangered species, you might hesitate to order spotted wobbegong in a restaurant. 'Roussette', the term used by William Anderson Smith, is one of several fancy names traditionally given to dogfish when it appears on a menu. 'Rock salmon' is another. In chippies in London and Kent you can order 'rock and chips', and never need to acknowledge that you're eating anything as unappealingly named as dogfish.

But how about mermaid's purses themselves? Is it possible to imagine eating them? Another mid-nineteenth-century natural history writer, the Reverend J. G. Wood, suggests it just might be.

In *The Common Objects of the Sea Shore; Including Hints for an Aquarium* (1857), he writes about skate egg cases (or 'skate-barrers' as the local fishermen call them, on account of their resemblance to handbarrows):

> I was once talking about these eggs to some fishermen, who told me that in the spring they often found these eggs before the young were hatched, and were accustomed to boil and eat them just as hens' eggs are eaten. Whether to believe them or not I could not make up my mind, for fishermen are wonderfully loose in their details. However as they gave me the information, I present it to the reader, and leave it to his own discretion to judge, or haply to his own energy to prove or disprove by actual experiment.

Thanks, Rev, I think I'll pass. But still I'll always be pleased to find a mermaid's purse, for its reminder of childhood, its hint of magic, its curious beauty, and also as a sign of hope for the future. Each discarded purse represents a new life. The ingenious design means that eggs, so vital but so fragile, are protected by their unique packaging, equipped either with elastic laces to attach it to seaweed, or with a specially roughened surface, like Velcro, to help it stay put until the moment of ripeness.

It's like a message in a bottle. The message is the DNA which will make the next generation. It's launched into the swift currents in its own disposable boat: light but tough, sewn shut against dangers, perfectly engineered for survival.

Prozac

A promising glint of silver in the tideline turns out to be a blister pack of Prozac. Its fourteen blisters, each labelled with a day of the week, are all empty. Like the egg case of some unidentified fish, it lies hatched and abandoned.

Blister packs are made partly from plastic, so it's at home here, among the bottle caps, chocolate wrappers and cotton-bud sticks. But it's even more problematic than most of this discarded plastic. In theory, much of the rubbish we find lying on beaches could be collected and taken to a recycling centre; but blister packs like this one are not recyclable. The plastic blister top is heat-bonded to the foil backing card, and the two are almost impossible to separate.

But that's just the packaging. The contents too have an environmental impact, even though this packet was obviously empty before it was thrown over the side of a ship or dropped by a dog walker. Traces of all the foods and medicines we consume end up, inevitably, in the ocean. We can't metabolise everything we swallow; the residue is excreted and flushed out and eventually finds its way into our rivers and seas. Some elements are very persistent; caffeine levels are so high in some coastal areas that it's used as a marker to determine general water quality.

Most of these residues are innocuous enough, but there are exceptions. In 2004, the Environment Agency raised the alarm about traces of fluoxetine in our drinking water. It was building up in rivers and groundwater supplies via the sewage system. Doctors issue thirty million prescriptions a year for Prozac and other anti-depressants classed as selective serotonin reuptake inhibitors. Fluoxetine, the active ingredient in these drugs, has been leaking into oceans and rivers for years, but the extraordinarily steep rise in prescription means that levels are rocketing.

The government quickly issued reassurances that the quantities were too diluted to matter, but not everyone agreed. 'This looks like a case of hidden mass medication upon the unsuspecting public,' said Norman Baker, then Environment Spokesman for the Liberal Democrats.

*

So have some of us now got Prozac 'on tap', whether we like it or not? It's difficult to find reliable information on how much fluox-etine is getting into water supplies, and whether it's enough to have a significant effect on those who drink it. But there are other potential dangers, which are certainly being documented and are starting to worry environmental scientists.

There's particular concern over the bizarre effects of antidepres-sants on sea life. For obvious reasons, the highest concentrations of fluoxetine are found in coastal areas, near towns and cities. The creatures who live in these coastal waters – especially shrimps – are right on the front line.

Fluoxetine works by changing levels of serotonin in the human brain, and now it's doing something similar to shrimp brains too. Serotonin is a neuro-hormone which can modulate mood and decrease anxiety. Of course, we have no way of knowing whether or not shrimps get depressed, but the serotonin effect kicks in anyway, and this can spell disaster for the shrimps. It alters their perceptions, making them more reckless than usual, and more vulnerable to unaccustomed danger. They start to behave in uncharacteristic ways. Instead of staying in the shadows, relatively safe from predators, they swim towards sunlit water, where they become easy prey and are picked off by passing fish and birds. The effect on shrimp populations in the most severely affected areas could be devastating, according to marine scientists.

That could be a painful blow here in my local area, where there are still remnants of a once thriving shrimping industry. Fifty years ago, there were scores of shrimpers working the Southport sands, following the ebb tide in their amphibious vehicles, trawling for brown shrimp. Now there is only a handful left, and a ramshackle collection of rusty vehicles abandoned at the top of the beach.

The decline in shrimp stocks has been blamed on a depressing litany of pollution problems: raw sewage, industrial chemicals, pesticides, radioactive effluent from Sellafield nuclear power station. It would take more than a few doses of Prozac to cheer up the people who once made a good honest living from these waters.

Potted shrimps are still produced here, but on a much smaller scale. They're made in the traditional way, cooked with butter and mace, and recently they've been picking up food awards and regaining some of their lost popularity. James Bond, that famous connoisseur, would undoubtedly approve.

Charles Simic, too. His poem 'Crazy About Her Shrimp' is a gleeful celebration of the combined physical pleasures of love, sex and food:

> No sooner have we made love
> Than we are back in the kitchen.
> While I chop the hot peppers,
> She wiggles her ass
> And stirs the shrimp on the stove.

How good the wine tastes
That has run red
Out of a laughing mouth!
Down her chin
And onto her naked tits.

'I'm getting fat,' she says,
Turning this way and that way
Before the mirror.
'I'm crazy about her shrimp!'
I shout to the gods above.

He sounds pretty loved-up. If he wants to stay that way, perhaps he'd better keep off the shrimp. Fluoxetine doesn't just enhance serotonin; it also suppresses dopamine; and dopamine gives us, amongst other things, the feelings of elation and exhilaration we experience when we fall in love.

The anthropologist Helen Fisher, while acknowledging the effectiveness of Prozac in treating clinical depression, is worried about its routine use in milder cases. 'We all know these drugs cripple your sex drive,' she says. 'But humanity has inherited other brain systems for reproduction as well, among them the neural mechanism for romantic love. And these men and women may be jeopardising this brain system too.'

We don't know whether the shrimp experiences any of these side effects. But there are reasons to be very worried about what's happening. Dr Alex Ford at the University of Portsmouth, who has been studying developments, says: 'crustaceans are crucial to the food chain and if shrimps' natural behaviour is being changed because of antidepressant levels in the sea this could seriously upset the natural balance of the ecosystem'.

Swim towards the light. Ironically, it sounds very much like something from a self-help manual for depression.

Gooseberries and Jelly

My son sees it first: a small spherical object on the sand, right on the water's edge. We crouch down and peer at it. It's cloudy and translucent, with a wisp of something inside. It reminds me of a particular pearly type of marble I used to collect as a child. In those days, I spent long hours with friends playing elaborate games of marbles in dents and grooves in the pavement, collecting my winnings in a Peek Frean's biscuit tin, and although it was the multicoloured 'beauts' you were meant to prize most highly, I secretly preferred those plainer translucent ones with their milky shimmer.

Just as I'm recalling this, another identical sphere rolls in on a gentle wave and comes to rest right next to the first.

Could they be plastic? I think of 'mermaid's tears', the droplets of waste from injection moulding factories which end up in the sea. These are much too big to be mermaid's tears, but they could well be plastic of some kind. I stand up and squint against the brightness to look southwards along the beach. Now I really look. And there are lots of them, dozens, hundreds, all along the edge of the sea, like a fringe of beads edging a shawl. Either there has been some serious pollution event here, or these are not plastic, but natural.

We start to speculate. They could be eggs. Various sorts of egg cases are washed up on the beach, most commonly mermaid's purse and another kind known as the sea wash ball which looks like an aggregated mass of dry brownish cells. But as soon as I touch one of the spheres, I know it's not an egg. It's cool and elastic to the touch. *Gelatinous.*

Back home, I consult a marine encyclopaedia and find it quickly. It's a comb jelly with the beautifully accurate name sea gooseberry. It can be almost perfectly spherical, or slightly ovoid, and just the size of a plump, ripe gooseberry. In life it has long, plumed tentacles which it uses to fish for plankton, but these must have been either withdrawn into their sheaths or detached when the ones I saw were cast onto the sand. Rachel Carson describes a swarm of comb jellies feeding on mackerel eggs, which are 'swept up in the silken meshes of the tentacles and carried by swift contraction to the waiting mouths'.

Comb jellies are biologically distinct from jellyfish, though the similarities are obvious. They are present in vast numbers in our coastal waters, and there are several species, including one with a flattened body like a pair of transparent wings, and another shaped like a silky ribbon, which swims with a sinuous wriggling motion and rejoices in the name of Venus's girdle.

The shimmering filaments I saw inside are swimming combs, each with transverse rows of hairs that beat in co-ordinated waves to propel it through the water. At night, these have a phosphorescent glow. When they swarm in shallow waters, they can be seen touching the waves with blue-white opalescence.

*

I have a photograph of myself with my daughter and her father on the beach. It must have been taken by our son, probably with his first camera. I know it was New Year's Day, and with that relatively reliable judgement of age we have about our children if about nothing else, I'd guess it might be 1998. Our faces are ruddy with cold, in spite of the fleece jackets, scarves and wellies. We're standing in a solemn little semicircle about the largest jellyfish I've ever seen in my life. It must be three feet in diameter.

We were all very curious about that jellyfish. I remember I looked it up when we got home, but identifying it was not an easy task. The problem is that descriptions and illustrations in books usually refer to the animal in its habitat: swimming, three-dimensional. In these pictures they display variations in shape and form and colour which are much less apparent when you find them lying on the beach. You can make out a few distinguishing features – radial lines, star shapes and other forms within, for instance – but there's no getting away from it, they are disappointing dead. The graceful, lucent creatures we see on film – expert swimmers, delicately frilled and glowing – are cruelly reduced in death to beached puddles of goo. The colour inside is dulled and nondescript against the sand. And the degradation doesn't stop there; they are often hacked by the beaks of seabirds. The ragged remains of a jelly is a common sight, surrounded by a havoc of bird footprints. They're even harder to recognise once they've been flipped over a few times and had chunks ripped out of them.

Eventually I identified the giant we saw on New Year's Day as *Rhizostoma octopus* or the barrel jelly. It's common enough, and sometimes known by the unflattering alternative name dustbin-lid jelly, but it's a very beautiful species in life. Its shape is reminiscent of a sky lantern. There's an ice-blue or tawny bell, and eight tentacles underneath, each with a mass of creamy frills which are actually multiple mouths that catch particles of food wafted in their direction. Even dead on the beach, it's solid enough to sit proud of the surface, so it can be seen from a little distance. Then it resembles the smooth dome of water that forms just before a geyser shoots into the air.

*

Jellyfish is a comic name. Jelly is wobbly and colourful and funny. In my mind it's forever associated with the annual Sunday School parties I attended as a child: surprisingly riotous affairs which always culminated in a jelly fight. The stuff came in three colours like traffic lights, and was dished up to us in waxed paper bowls, along with a spoonful of that vanished delicacy, blancmange. The plastic spoon had just the right degree of flexibility to bend back and catapult a blob of jelly from one side of the trestle table to the other. There would be a satisfying splat as it found its distant target on the back of a boy's neck.

At the same time, jelly can have less pleasant associations. Petroleum jelly, with its industrial smell and extreme greasiness. The jelly in a pork pie (some people's favourite bit, allegedly). Calf's foot jelly, which is no doubt fantastically nutritious. In my mother's old *Radiation Cookbook* there was a chapter on 'Invalid Cookery', and my brothers and I would torment each other by reading out the recipe for calf's foot jelly, along with other choice items such as brain fritters and veal tea. Why such queasy-sounding items should be considered good choices for the delicate stomach is beyond me.

Jellied eels are another acquired taste. In his book *Waterlog*, Roger Deakin recalls being taken out eel-trapping on the Norfolk

Broads, and finds out how proper jellying is done: you should never add gelatine, as some inferior producers do, but simply boil the eels in their skins, since that's where the real jelly is naturally present. But the really off-putting thing is their diet: let's just say they are very unfussy eaters, and not at all averse to scavenging a drowned corpse, should there happen to be one available.

Jelly, with all its funny and stomach-turning associations, helps to define the way we think about jellyfish. These are slippery creatures, wobbly and elusive. Sometimes, visiting Antony Gormley's *Another Place* at Crosby, you can see them left behind by the receding tide, resting like wet berets on top of the heads of the iron men. They're alien, made of stuff which could hardly be more different from our own solid flesh, and so insubstantial you can see right through them. They have comic roles in children's books and films, but the thought of brushing against one while swimming is repulsive enough to keep some people out of the water even on the hottest day.

*

In his famous essay of the same title, the philosopher Thomas Nagel investigated the mind/body problem by posing the question 'What is it like to be a bat?' His conclusion was that we can never access, even through the imagination, the subjective experience of a creature so fundamentally unlike ourselves. 'I have chosen bats instead of wasps or flounders,' he explained, 'because if one travels too far down the phylogenetic tree, people gradually shed their faith that there is experience there at all.'

Flounders are a telling example; the lives of marine creatures are generally quite mysterious to us. Leigh Hunt's 1836 poem 'The Fish, the Man and the Spirit' is a dialogue between disparate and seemingly irreconcilable beings. It opens with a man addressing a fish:

> You strange, astonish'd-looking, angle-faced
> >Dreary-mouth'd, gaping wretches of the sea,

and concludes with the fish-spirit attempting to summarise the essential differences between them:

> Man's life is warm, glad, sad, 'twixt loves and graves,
> >Boundless in hope, honour'd with pangs austere,
> Heaven-gazing; and his angel-wings he craves:
> >The fish is swift, small-needing, vague yet clear,
> A cold, sweet, silver life, wrapp'd in round waves,
> >Quicken'd with touches of transporting fear.

So what can it possibly be like to be a jellyfish? How much harder we have to strain the imagination to conceive of it. They are in every way unknowable, far more remote from human experience than fish. More remote even than, say, seahorses, which have very strange lifestyles indeed but at least have a hint of the mammaloid about them.

Jellyfish have no skeleton, no circulation, no digestive organs, no nervous system. Sandwiched between their two layers of skin is a gelatinous substance called mesoglea, which gives them structural support in water. They need no respiratory system, because their skin is thin enough to allow oxygen in by diffusion. And what is social interaction like if you don't have a face? Some kinds do have rudimentary eyes, but they are not capable of forming images, only of sensing light.

But something more than the material and experiential gulf between us makes us cautious about jellyfish. Everyone knows that they sting. They have a primitive sense of touch and movement by means of a 'nerve net', a loose network of nerves in the skin, and when this is activated, whether by a hungry shark or a hapless swimmer, they strike. The tentacles are armed with stinging cells, or nematocysts – small capsules with a barbed thread coiled within. The stinging cells can be sprung in a fraction of a second, triggered by the need to defend or attack.

Most jellyfish sting rarely, mildly, or not at all; but naturally it's the exceptions that capture the public imagination. One dreaded stinging species, the Portuguese man-of-war, is habitually an oceanic creature, but is sometimes driven onshore by south-west winds. Its sting has been known to kill by causing an anaphylactic reaction, an overwhelming allergic response which can leave the victim unable to move or breathe. But in the vast majority of cases its sting is painful rather than life-threatening; victims have compared it to the lash of a whip.

A few exotic species, so far strangers to British shores, are genuinely dangerous. The most notorious group is the box jellyfish, which has caused many deaths in the Philippines, Malaysia, Australia and Japan. One of the largest species, known as the sea wasp, carries enough venom in its body to kill sixty people. It's about the size of a basketball, and it's said to have markings which give it an eerie resemblance to a human skull. But you'd be unlikely to notice; it's transparent, and rarely seen. By the time you know it's there, it's too late.

At the other end of the scale, a species about the size of a peanut is so minuscule and inoffensive-looking that it's virtually impossible for swimmers to avoid. If touched, it fires millions of microscopic nematocysts, not just from its tentacles but from all over its body, delivering a tiny but life-threatening payload of toxin. Victims enter a notorious state of poisoning known as Irukandji Syndrome, characterised by extreme pain throughout the body, racing heart, rocketing blood pressure and an overwhelming feeling of impending doom. Patients are sometimes so certain they are going to die that they beg their doctors to kill them to get it over with.

Irukandji Syndrome was something of a mystery until 1964, when an Australian marine biologist, Dr Jack Barnes, determined to prove his own hypothesis, spent several hours lying in the water in a wetsuit, waiting to see one of the thumbnail-sized jellyfish he suspected of being the cause. When one swam past his mask, he grabbed it. He then deliberately stung not only himself but also his nine-year-old son and a lifeguard. All three were rushed to hospital with Irukandji Syndrome; the point was proven. Does this speak of heroism, or reckless exploitation, or both? It was a scientific breakthrough; but it remains an ethically troubling story, like the one of Edward Jenner injecting the gardener's son with cowpox, or John Gummer feeding his daughter a beefburger on live television at the height of the BSE crisis.

*

Divers and surfers are especially prone to jellyfish attack, and surfing websites are full of conflicting advice on the subject. First aid has to be administered quickly to deactivate and then remove the nematocysts attached to the victim's skin. But there's confusion and disagreement on the best way of going about it. It's easy to do more harm than good – clumsy handling can trigger any unfired nematocysts and pump more venom into the victim, or even the rescuer.

For the first stage – deactivation of the sting – one of the most common recommendations is the use of a meat tenderiser: not the wooden kitchen implement, but the stuff you can buy in a jar in the supermarket. Some experts dismiss this as a waste of time. Vinegar is said to be beneficial in some cases, but disastrous in others. Alcohol is often applied, but makes matters worse. Urine has a long history as a folk remedy – and I'm told there's an episode of *Friends* in which Monica is stung and treated in this way, so it's clearly well established in the popular imagination – but it seems it may be no use at all.

There's more of a consensus about the next step, the removal of the stinging cells. It requires a tool: preferably the edge of a sharp knife, but in *extremis* a credit card. (Reading this, it occurs to me that the credit card is a surprisingly versatile object: during my university days, women students were advised to cut an old one in half, and carry it in a pocket to use in self-defence; more recently, I've pressed them into service to scrape ice off the windscreen and even to break into my own house.)

All sources agree that before coming to the rescue of a jellyfish casualty you should put on protective clothing of some kind. If you're on the beach, you're probably wearing less than usual, and exposed skin must be covered unless you too are to be affected by the venom. You'll have to improvise. 'Put on whatever protective clothing is available,' instructs one website, 'such as gloves, wetsuit, or pantyhose'. Pantyhose may not be easy to procure in a hurry. But if you can manage it, and assuming the victim survives both the attack and the ministrations that follow, he could be in for a shock when he opens his eyes and turns to thank his rescuer.

There's little danger from the jellyfish swimming in the waters off this shore. The barrel jelly is harmless, and so is another common local species, the moon jelly. Others I've seen here – the compass and lion's mane jellyfish – can deliver something similar to a wasp sting, particularly unpleasant if it catches a sensitive area like the face or the genitals.

But Irukandji could be coming to a coast near you. Scientists are warning that jellyfish are on the move. Populations are exploding, and spreading beyond their usual bounds. It's partly because some of their natural predators, such as leatherback turtles, have seen a catastrophic decline in numbers in recent years, leaving jellyfish less and less vulnerable to being eaten. But it's also down to their exceptional ability to adapt and survive. Because they're such hardy creatures, they continue to thrive even in places where overfishing, pollution and rising sea temperatures have killed off fish, corals and other species. No matter how impoverished conditions become in the dead zones of our oceans, they just keep on multiplying. As biodiversity is stripped away, the jellyfish shall inherit the waters.

Their soft, invertebrate bodies rarely leave a fossil record. But they are very ancient creatures. One exceptional fossil discovery in Utah dates to five hundred million years ago, during the period known as the Cambrian Explosion, when a huge diversity of complex organisms developed in a relatively short time. (This was an 'explosion' in slow motion – it took several million years, but that's fast by evolutionary standards.) Jellyfish, sea anemones and corals are amongst the countless species that emerged during this time.

The Utah fossils record a swarm of jellyfish, but swarming is still poorly understood today. Research is underway to determine what causes it. One project in the Irish Sea is tracking and monitoring the numbers and behaviour of local species. Amazingly, the researchers have actually found a way of tagging individual jellyfish, and photographing them from low-flying aircraft, in order to work out how and why they congregate in such large numbers in specific areas and at specific times. Swarms or 'blooms' consisting of hundreds of thousands of individuals seem to occur in response to seasonal changes, ocean currents and the availability of food. The phases of the moon also play a role: on the coast of Hawaii, box jellyfish turn up regularly at the same point in each lunar cycle, and stay there during the period from eight to twelve days after a full moon. No one knows why they suddenly move on, or where they go next.

It's clear that swarming is a normal part of jellyfish life. Now, though, there are new and ominous developments. Vast numbers have been overwhelming fish farms and destroying everything. In 2007 Northern Ireland's only organic salmon farm lost its entire stock this way when a sudden influx of mauve stingers flooded the cages, stinging and suffocating the fish. Workers in boats tried desperately to intervene, but the sea was so thick with jellyfish that the boats' propellers became jammed and they couldn't get there in time. Reports tell of vast areas of the Gulf of Mexico becoming completely choked with jellyfish, to the exclusion of everything else. Photographs show Japanese fishermen hauling in huge nets of them instead of fish. They've even been blamed for causing electricity blackouts in the Philippines, by clogging up the cooling pumps of a power station.

*

'If we do not change our ways we will have less and less to catch . . . so jellyfish could end up on the menu as opposed to cod in our fish and chips,' says Professor Callum Roberts, a marine biologist at the University of York.

Eating jellyfish is by no means a new idea. The barrel jelly is one of several species considered a delicacy in China. There are jellyfish masters, who specialise in processing them in time-honoured traditional fashion: salting, compressing and dehy-drating the jelly until it's creamy-coloured and crisp in texture. The processed product is then soaked in water overnight and eaten cooked or raw, dressed in oil, soy sauce and vinegar.

Perhaps more of us are going to have to start thinking of them as a resource, not only as food but in other applications too. Already, a pharmaceutical company in the USA claims to be 'breaking new ground in protecting brain cells against neuronal degeneration' through 'cutting-edge applications of the jellyfish calcium-binding protein apoaequorin'. In other words: jellyfish pills. They're being hailed by some as a wonder treatment for Alzheimer's disease and other kinds of dementia. Apoaequorin is the protein which gives some jellyfish their luminescent glow, and it's said to improve memory function, though not all scientists are convinced.

Humans are not the only ones to find ways of making use of the plentiful supply of jellyfish. Some specialised organisms live in symbiosis with them, or even right inside them. A tiny crustacean called Hyperia galba spends its entire adult life this way. It lives off the mass of jellyfish eggs and particles of food consumed by its host. The seabirds I've observed pecking at beached jellyfish were probably searching for these nutritious little hitchhikers. I saw one once, still alive inside the transparent body of a dead moon jelly. It was a pale and insignificant creature, except for its green eyes like specks of sea-glass.

*

The lion's mane is a regular in the Irish Sea. Washed up on the beach it still looks dramatic: a golden-brown eight-petalled flower shape in the centre, surrounded by a blood-red ruff. Another frequent find is the compass jelly, the colour of straw and marked with a circle of V-shaped points. These are two of only five or six recognisably different kinds I've ever seen here.

The mauve stinger, so notorious in the salmon farming industry, is just one of a number of 'alien' species said to have 'invaded' British shores in the last twenty years. The vocabulary is reminiscent of tabloid scare stories about immigrants and asylum seekers. Our own lovable jellyfish rogues are one thing, but look out! – dangerous foreigners are breaching our defences, and everything is going to hell in a handcart.

Populations have never been static; the history of life on earth has been one of changing conditions, and animals, like people, have adapted to change by moving and colonising new territories. But against this background of longer-term flexibility it is certain that human activity is now bringing about fast and dramatic change. Our appalling management of the oceans is leaving ecological gaps for jellyfish to move in, and making new areas of the ocean habitable for species which could not previously have tolerated conditions there.

A single exotic species found on the beach may be a portent
. . . providing you can identify it. With that thought, I'm going
back to the reference books. Those of us with beaches to walk
on should be learning the language of the things we find there.
We should be reading the signs.

The Underworld

Spring tides have spring-cleaned the beach. The rubble of the tide-line – which was heaped high when I was here last week – has all gone, swept back into the sea to be washed and mixed and sorted again. For those castoffs, the journey is never over, and there is no such thing as home.

At the very top of the beach, there's a neat kerb cut into the foot of the dunes by the last tide. The sand is smooth and clean as a polished floor. Three or four horseboxes are parked near the beach entrance, and racehorses from Aintree are exercising, their galloping hooves thundering on the new-made surface.

As the tide recedes, buoys are exposed, most of them home-made out of old plastic drums or petrol cans. Each is tethered by a fraying length of orange or green plastic rope to a stake hammered into the sand in the intertidal zone.

Between and around the buoys are tiny replicas, bright green floats lying stranded, some broken loose, others still moored to the sand by a long thread.

These are egg masses laid by paddleworms. The eggs are more frequently seen than the worms themselves, which live under stones or in muddy sand at low water. They lay their eggs in these rounded masses, anchoring them either to seaweed or in sand. Amongst weed they're well camouflaged, but exposed on the beach they're very noticeable. They dot the shining surface, almost luminous in their greenness, like blobs of alien goo spilt from a low-flying spaceship.

There are several distinct species of paddleworm, and paddleworms are just one in a host of families of worm species which inhabit the beach. Flatworms, ribbon worms, sludge worms, bristle worms, cat worms, scale worms, fan worms, tube worms – there are hundreds of species, millions of individuals, burrowing under the surfaces of our beaches, riddling the mud with holes and tunnels. Theirs is the secret life of the beach; sand may look inert, but in fact it's seething with industry. We can only guess at the detail of that hidden world as we crouch and examine the scribble of wormcasts along the water's edge.

Perhaps the best-known of marine worms is the lugworm. At low tide there is often a fisherman or two, with bucket and waders, digging furiously. He waits for a wave to wash in, and as it pulls back he heels the spade into the wet sand and digs away as if his life depended on it. It looks manic behaviour, puzzling to the uninitiated. I followed one man to try and get a closer look at what he was doing, but no sooner had he dug a hole and chucked something in the bucket than he was off, moving a few yards along, digging again, moving on. It was like watching an opportunist thief trying the door handles of parked cars, testing and passing on to the next, quick and stealthy.

Lugworms live in U-shaped burrows under the sand. If you want to dig them out, you have to look for the characteristic cast on the surface, and for a tiny blowhole or feeding depression nearby. Then you dig between the two with your spade, and hope for the best. A few brisk cuts of sand should bring the worm to the surface, and you pick it up and throw it into your bucket.

It's hard physical work, but serious anglers swear by it. These are dedicated hunters, going after the bait with the same energy and determination they show in pursuing the prey itself. These devotees don't think much of the dried or frozen worms you can buy, ready-packaged. Fresh lug is the real thing: that's what really fetches the fish, especially the cod.

For many years anglers have distinguished between two types of lugworm, but it's only relatively recently that biologists have caught up with them and confirmed that there are indeed two species. They're subtly different – one is slightly larger and pinker, one a little more bristly. Their lifestyles are essentially similar: they eat their way through the muddy sand, digesting what they can and excreting the rest. Their casts are studded with pale fragments of shell. Blow lug, or common lug, is the one most often dug and used for bait. Black lug is highly prized, but more difficult to dig for as it burrows deeper and doesn't usually make a blowhole. To catch black lug, you really need a bait pump, which you work by hand to suck up large quantities of wet sand and – with luck – your quarry.

Once you've collected your bait, the next challenge is to keep it alive for as long as possible. A photograph in an angling magazine shows one recommended method: the worms are spread in an old cat litter tray, layered with newspaper, and kept in the fridge. Other, more conventional refrigerated goods are clearly visible: a pot of yogurt, a tub of margarine. I imagine this practice may not be very popular with other members of the household. The next photograph, in which a live worm is threaded on a baiting needle, oozing blood onto a kitchen chopping board, could well constitute evidence in the divorce courts.

Cod may love them, but they're not the only ones. Seabirds feast on marine worms, just as blackbirds eat earthworms in the garden.

Along with cockles, clams and mussels, they are a staple of the oystercatcher's diet, and sandpipers are partial to them too. Gulls have famously varied tastes, happy to scavenge dead fish and crabs on the beach as well as hoovering up chips and bits of old sandwich left by picnickers, but they are very fond of worms and other invertebrates. And given the variety of species living here, they can afford to be choosy.

Some worms can be spotted easily, especially on shores where there's more shelter in the way of rocks or seaweed. They range from a thin red type like the liquorice laces I used to buy from the sweet shop on the way to school, to a short, squat variety, scaled and warty, which rolls up like a woodlouse if you try and touch it.

But most of them we'll never see, except in the pages of reference books. Their ways are closed to us, their true elements mud and water rather than air. Their visible signs are the delicate structures they leave on the surface – the neatly coiled cast of the black lug, the raggedly frilled tube of the sand mason, or – most spectacularly – the cellular colonies built by the honeycomb worm, which I've seen on rocky shores in Wales and Dorset. It's not something you would necessarily associate with worms; it's on a larger scale, made up of thousands of individual tubes, constructed from sand cemented together into a hard, rock-like substance like a kind of reef.

The Old English word *wyrm*, or serpent, was also used to mean 'dragon', and fearsome *wyrms* terrorised people from Lambton to Cornwall. One of them, known as the *knucker* or *nicor*, was a water-dwelling species; in *Beowulf* there are references to these creatures swimming in the sea and basking on the cliffs as Beowulf passes by on his way to the mere to confront Grendel. Their relatives in the world of marine worms may be very much smaller, though it wasn't always so: in central Spain, in an area which was once seabed, fossils have been discovered of giant marine worms, a metre long and fifteen centimetres in diameter. But seen through a microscope even some of our present-day species look quite monstrous. Some are armed with spines, hooks and horns, some have tentacles, others multiple pairs of eyes. If, like Alice, I drank a potion and shrank, small enough to slip not down a rabbit-hole but into the *nicor-hus* of one of these creatures, I'd be in for a pretty terrifying experience.

Like a cosmic wormhole, it might feel like a short cut through spacetime, whisking me faster than the speed of light from one reality to another. And that's probably how it feels to be a lugworm: dug violently from your lair and thrown in a bucket, your only future the fridge, the baiting needle and the spear-sharp teeth of a cod.

Black Gold

Some of the beaches I remember most vividly from childhood were of thrillingly rough and varied terrain, with caves, and rocks to climb and cut your shins on, and rockpools full of alien creatures with frills and pincers that scuttled or shifted shape when you dipped your bucket in. My brothers and I would go off on journeys of discovery which took all afternoon, over rocks slithery with damp seaweed, where it grew thickly enough to trap your foot.

Equally memorable are the swimming expeditions when we waded in, quaking and goosebumped – with the cold, but also with the thrill of the unknown, of what might be lurking under that restless surface. In a moment of bravado I would launch into the doggy-paddle, start to enjoy the freshness of the water, feel the sting of the salt on scratches and sunburn. Then suddenly I'd be snared and squealing, booby-trapped in a mass of seaweed like tangled wires. One of us would tear out a length like a whip and chase the others out of the sea and up the beach, and the fight that broke out ended, every time, with sand in someone's eye.

Back then, I knew seaweed in that intensely physical way in which children know the world. It was slippery and dangerous, you could get caught in it and almost drown, it hurt when it hit you in the face.

Adults can feel ambivalent about it too. It may have become an ingredient in expensive cosmetics and dietary supplements, but it still has the 'shudder factor'. It still traps and ensnares. There are stories of ships found floating abandoned in the Sargasso Sea, after sailors becalmed in this notoriously windless region were driven mad by thirst and went overboard to try and walk across the 'fields' of seaweed that surrounded them.

Yet seaweeds have their attractions. They are beautiful, especially when seen waving gracefully underwater rather than limp on the beach. They are mysterious too: relatively little is known about them, and there's still a degree of confusion on the fundamental question of what they are. The dictionary may define them as 'simple aquatic plants', but biologists disagree. In spite of appearances, they don't have roots, leaves, branches or fruit, and are only distantly related to land plants. In fact, they're classified as algae: a staggeringly large and diverse group of species, ranging from simple single-celled organisms to giant kelp sixty-five metres long. Some algae thrive in water, others on land, and they are even found growing on snow and ice. Taxonomy is a surprisingly controversial business, and algae, along with other equivocal organisms like parasites, amoebas and moulds, have a history of drifting like refugees between categories, setting up camp in one only to be moved on to another.

Seaweeds are slippery customers. They exist in unimaginably huge quantities, more or less unmapped and unexamined. About ten thousand species are known so far, but new ones are being identified all the time. They may look roughly alike when we see them cast on the beach, but in fact they are remarkably various.

★

My local beach is not a seaweed beach. There is no live seaweed growing here; there's nothing for it to grow on. There are no cliffs or rocks or caves, hardly any landmarks at all. It's all surface and distance and reflection.

Nevertheless, there is seaweed to be found in modest quantities, if you explore the narrow ribbon of debris known as the wrackline. Wrack or weed, brought in on the tide and cast on the sand, shows the extent of high and low tide. The line at the top of the beach, bleached and desiccated, marks the limit of the last storm tide; and this one, nearest the sea, is where the most recent tide reached, just a couple of hours ago. This line is sparse, its debris relatively fresh, and one of the major constituents is seaweed. At first glance, wreathed together amongst bits of wood and rubbish and dead crabs, it all looks the same. But as soon as I start picking it over I can distinguish several different kinds.

The first piece I pick up is a thick, olive-green strap, leathery in texture, with little stalks sporting fat dark fruits the size of grapes. *Bladderwrack*, I say automatically. Appearances are deceptive: these are not fruits, but blisters or gas bladders which act like miniature balloons, keeping it afloat.

But if I've got that one right, what about this next one: a long, branched frond with bladders at intervals all along it, smaller and more spherical, the size of raisins rather than grapes?

And this: darker, spiralled, with swollen tips, each in the shape of a heart?

Back home, I tip out my bag and spread out my finds on the paving slabs. They are gritty and rubbery to the touch. With the help of my *Collins Pocket Guide to the Seashore of Britain and Europe*, I try to identify what I've got. I'm pretty sure of four separate types of wrack. The one I found first and felt so confident about is actually not bladderwrack, but egg wrack (sometimes known as knotted wrack). Bladderwrack is the one with its bladders on the frond rather than on separate stalks; the book gives me the alternative name popweed, which evokes another familiar childhood pastime, a bit of fun I recall every time I open a parcel packed with bubble wrap. The third kind, the one with the hearts, is spiral wrack or flat wrack, and I've now disentangled a fourth kind, channelled wrack, such a dark shade of green it looks almost black in the late afternoon sun, and with curled fronds that form a kind of groove or gutter.

I also have something quite different, not a wrack at all: a bright green tuft of fine threads, almost like moss, which may be Cladophora rupestris, or Ectocarpus, or Enteromorpha, or something else again – there are too many possibilities which look almost identical to my inexpert eye. Tangled up in it are a couple of stems of something I recognise as lyme grass, not a seaweed at all but one of the 'early colonisers', the first plants to take hold at the top of the beach.

Four types of wrack: not bad. If only I could have found toothed wrack and sea oak too. Or any of the other sixty or so varieties of seaweed listed in the Collins Guide, with their beautiful and enigmatic names: gutweed, sea lettuce, velvet horn, dabberlocks, furbelows, thongweed, peacock's tail, sea noodle, red rags, pepper dulse. Some names, such as mermaid's tresses and sea mare's tail, are descriptive of the visual appearance of a species. Others are keys to surprising little narratives about their characteristics ('the hollow sphere of the oyster thief, when filled with air, may lift a light object off the bottom so that it floats away'); or the human uses to which they have been put ('the poor man's weather glass, when hung in air, becomes soft and limp in humid conditions').

*

It's difficult to look at seaweeds without thinking of branches, stems and leaves, but appearances are deceptive. Instead of being rooted in the ground, they attach themselves to buried rocks by means of a kind of sucker called the 'holdfast'. They photo-synthesise, as plants do, but there are obvious limitations on photosynthesis underwater, where there is much less available light than on land. Sunlight is lost at the surface through reflection and refraction; just four or five metres down all red light becomes absorbed; and even blue-green light can't penetrate deeper than fifty metres or so. For this reason, most seaweed occurs only in shallow waters, in the 'photic zone'. In disturbed waters near the coast, suspended particles of stuff stirred up from the sea bottom make the available light so scattered that seaweed and other algae are even more restricted.

Another reason why it's common to think of seaweeds as plants is that they have a long history as human food. Ironically, although Britain is an island nation, the first examples that come to mind are Japanese varieties: nori, the thin, dark-green sheets wrapped around sushi; and wakame soup on the shelves of health food shops. A friend gave me some wakame a few years ago, and although I was sure I would like it – I pride myself on liking almost every kind of food – I was retching unbecomingly into her kitchen sink after the first sip. To my tastebuds it was just a mouthful of warm seawater, and it triggered a minor panic by reminding me of all the flounderings and near-drownings of my childhood trips to the seaside. Soon afterwards, another friend came to a party bringing a paper bag full of pieces of dried seaweed. He buys it in a Japanese supermarket and eats it like crisps. I thanked him politely for the gift, but couldn't help feeling that a bottle of Sauvignon would have done just as well.

In Japan, seaweed takes many forms. It's sold fresh, processed into powder, preserved in sugar, canned. You can buy it in strips or sheets which you soak in water to reconstitute it before adding it to sauces and broths. It's such a staple of the Far Eastern diet that the average person consumes nearly four kilograms a year. Demand on this scale can't be supplied by harvesting it from the wild alone, and seaweed cultivation in Japan and China goes back at least as far as the eighteenth century. At first nori was scraped from dock pilings, or seeded on ropes, or on stones placed on the sea bottom. Later, areas of rocky shore were specially blasted to create conditions in which it could be grown in larger quantities. These days, cultivation is big business, and it takes place on an industrial scale, on floating artificial structures built in inland tanks pumped full of seawater.

If, as Robin Cook announced a decade ago, chicken tikka masala is a true British national dish, futomaki may not be far behind. For now there's still a lingering whiff of the exotic about edible seaweed. But not very long ago, it was local and common-place. It was home-produced, and as familiar to coastal people as blackberries or cabbage.

An oil painting by William Collins, a landscape artist working around the turn of the nineteenth century, shows two young chil-dren gathering seaweed into a basket. Collins was fascinated by the seashore, and returned again and again, often painting people at work there – fisherwomen, shrimp boys, prawn catchers. He depicted the coast primarily as a place of labour rather than leisure. In the painting, the children work barefoot against a backdrop of rugged cliffs and a cottage with a smoking chimney. Wrack hangs from their arms in great, loose chains, black as coal. It's a picture of poverty, but also of plenty. Here was a crop that belonged to no lord of the manor; the shore was common land and what grew there could be taken freely by anyone. Seaweed grew in abundance on our shores, and so without question it was harvested and used. What could be made palatable was eaten, and the rest was put to other uses. It would have been unthinkable to neglect such a readily available resource; people simply could not afford to do so. It's only very recently that we've become prosperous enough to let things go to waste, and one of the consequences is a feeling of alienation from the idea of wild food.

But perhaps it's less alien than we think. In fact we all eat seaweed, mostly without knowing it. For years I've been buying jelly crystals for my vegetarian son. When I investigated the list on the packet, I found that the principal ingredient is carrageenan. A little research tells me that this lovely word, from the Irish *carraigín*, describes a substance extracted from 'Irish moss', which is actually not moss at all but a type of red seaweed.

Carrageenan is one in a family of substances known as hydrocolloids, which form a gel when they come into contact with water. This characteristic makes them extremely useful in processed foods. You can find them listed on packaging, in the guise of additives such as emulsifiers, thickeners, stabilisers and gelling agents. We eat them in ice cream, dairy products, processed meats and cheeses, jellies and confectionery. One of their more dubious uses is in cheap jam, where they sneakily bind together fruit purée into blobs which look and feel a bit like actual strawberries and blackcurrants. And perhaps their most intriguing manifestation is in Irish Moss, a Jamaican drink, sweetened and flavoured with rum and spices, which is known as an aphrodisiac and a cure for male impotence.

All this is a long way from the children scooping wrack into their basket. We may eat nori and wakame flown in from Japan, we may spread hydrocolloids on our toast, but few of us have tasted what grows on our own local shores. We have forgotten our own treasure.

*

Previous experience with the wakame suggests that I am not a natural connoisseur of seaweeds, but I'm curious. To the educated palate they are said to represent a rich and varied range of flavours. There are a number of types considered edible, including dulse, sea lettuce and oarweed, each with its own distinct flavour, like the different taste qualities of grapes in wine. But the only British seaweed harvested for food on any significant scale these days is purple laver, which is essentially similar to the famous Japanese nori. It has a broad frond a bit like cabbage leaf in appearance but thinner; it's green when young but becomes purplish-red in maturity, and almost black when dried out. The only surviving culinary use is *bara lawr* or laverbread, which is still made in parts of Wales. It isn't bread at all, but a sort of savoury pudding of boiled seaweed. After harvesting, the laver is washed, boiled for five or six hours, then minced. It's often fried and eaten with fish or bacon. Alternatively it can be spread on toast and sprinkled with vinegar. A version of laverbread used to be made in parts of Scotland too, where laver was known as sloke and was harvested and processed as a matter of course by crofters.

There are rumours of renewed interest in laverbread amongst a younger generation of chefs, and it's begun to make an appearance on the menu of upmarket restaurants. Food revivals have a habit of taking what used to be ordinary and basic and made in the home, and turning it into something expensive for the gourmet market (think wild garlic, or oysters). Fashions come and go quickly, and it's easy enough to be cynical about these reinventions. But perhaps something genuine and good is going on here: the rediscovery of the local, of the home-grown. A revived awareness of the richness all around us. A desire to know where things come from before they end up on our plates.

Either way, I've decided to fly in the face of personal experience, and to see whether I can acquire a taste for laverbread. Aficionados go into raptures about it; sometimes it's described as 'black gold', and Richard Burton once called it 'the Welshman's caviar'. It has its own official website, where its history, origins and nutritional benefits are explored, and a recipe page suggests a surprising range of dishes, including 'seared tuna with laverbread salsa', 'Chinese crab, egg and laver soup' and the intriguingly named 'laverbread and pantysgawn quiche', which turns out to be made with an award-winning soft goat's cheese from Monmouthshire.

A series of emails and phone calls has confirmed what I suspected: that proper tasting will require a field trip, since the tinned variety which can be ordered by post is not quite the real thing. I will have to go to the source. The undisputed centre of laverbread production is South Wales, and the place to buy it is the seafood rotunda on Swansea Market. I call my partner, and somehow manage to convince him that a trip there would make a good Valentine's weekend away. There's to be a rugby international on the day itself, which probably rules out a romantic dinner for two anywhere in South Wales; but still, I assure him, it will be an experience.

Swansea has a large and thriving indoor market, one of those which has hung on to its role at the heart of the city's life. The Cockle Rotunda stands in pride of place right at the centre, at the very hub of activity. The other stalls ranged around it are emblazoned with names which seem to reach back decades, before the corporate anonymity of the high street and the cuteness of the smart suburban delicatessen, right back to an age of small family shops you might think had vanished for ever: *Billy Upton Butcher, Holland's Nets, Taffy's Barbers, Tom Whitehouse Electricals, Mair Harries Welsh Produce.*

On the rotunda, *Carol Watts Gower Cockles* is stocked with trays of lobster, jellied eels, mussels, whelks, fresh cockles in their shells, and polystyrene tubs of the Black Gold itself: 200g for £1.70, 500g for £3.50. It's for sale in sealed plastic pouches too, dusted with pale stuff Carol tells me is oatmeal. It binds the laverbread together, she explains, so that you can shape it into cakes and fry it up with breakfast. Indeed, I've already noticed it listed on the 'Welsh breakfast' menu at a neighbouring stall, along with bacon, eggs, tomatoes, black pudding, mushrooms, fried bread and hash browns (which, admittedly, had less of a ring of Welsh authenticity). Carol also sells it in tins, packaged complete with the legend *A Traditional Welsh Delicacy* in Celtic lettering and a symbol of approval by the Vegetarian Society. When I ask her whether it's as good as the fresh, she smiles and says diplomatically: Almost.

The illustration on the package shows what looks oddly like a Chinese willow-pattern bowl full of a gleaming black substance, garnished with tomato, green pepper and parsley, and with an ornate silver spoon balanced at an inviting angle. We buy a tin to take home, and a small tub of the fresh version to eat right away. The polystyrene tub is handed over, and we stand poking the stuff with our plastic forks for a minute. I smile bravely, trying not to breathe in, or to think of the wakame soup.

It's mushy in texture, very nearly black but with a green tinge.

It smells of the sea. I dig in and taste my first laverbread. It's cold, salty, soft in texture. Three or four forkfuls is enough. Still, I'm pleased to have passed this first test. Now Carol is saying that in her view it's better hot, spread on toast. It's clear this is just the start of my gastronomic adventure.

Laverbread's spiritual home is the Gower peninsula: the thumb of land sticking out under the hand of south-west Wales. The Gower is just twenty-five miles long, but it's a popular summer destination and a magnet for surfers. On this cold February day, there is just a scattering of the most determined of them on Rhossili beach, their voices torn to rags and blown away by the wind. It's not exactly California. It's parking at the campsite, all boarded up for the winter. It's changing into your wetsuit in the car park, then sprinting out onto the windswept sands with your board.

This doesn't look like seaweed territory to me: all golden sands and low dunes, not a rock in sight. We drive on round the tip of the peninsula, and the picture-postcard beaches along the western edge give way first to a smudged-charcoal landscape of tidal estuary and saltmarsh on the north coast, then to rocky coves sheltering under limestone cliffs to the south. These two environments are the source of two celebrated local harvests: cockles from the estuary, and laver from the rocky southern shore.

I'm not sure what I was expecting, but I'm surprised to trace this morning's laverbread to a functional factory building beside the estuary. There's a factory shop, which sells cockles too, but it's shut on Saturdays. Beside the road there are dozens of gigantic builder's sacks, all heaped full of cockle shells. The ground is covered with shells too, and there's a hopper-and-chute contraption for loading them into the sacks. Apparently there's a trade in cockle shells for decorative landscaping; they're used for laying paths and driveways, and crushed they make a snail-repellent mulch for the garden. It seems Mary Mary Quite Contrary was on to something.

A few miles further up the coast, we stop for lunch at a cosy pub in Llanrhidian. We have the place to ourselves apart from Mike, the landlord. Yes, we're doing food, he says, pointing to the blackboard, which features not one but two laverbread dishes: Laverbread with Cockles, and Laverbread Bake. He recommends the Bake: even people who don't think they'll like laverbread like the Bake. It's a piping hot mixture of laverbread, cockles, bacon and onion, topped with melted cheese, and triangles of crisp buttered toast on the side. You could almost think you were eating spinach, but for a subtle seaside aftertaste. I decide this is how I like it: hot, and combined with other flavours. Perhaps you have to be South Wales born and bred to enjoy it cold and straight.

I tell Mike I'd like to see some laver growing, but he gives me an enigmatic smile and says you have to know what you're looking at. He warns me, somewhat unnecessarily, against harvesting and cooking it myself: apparently it's impossible to wash all the sand off by hand, no matter how hard you try. Without the special machines they have at the factory, it stays forever gritty. Then he lets me into a secret. The laver, he tells me, is mostly grown in Ireland and Scotland these days, and brought here in articulated lorries to be processed. It's only South Wales where the tradition lives on, he says with pride. They don't value the stuff in those other places, they just scrape it up and freight it over here.

*

We are learning, belatedly, about the interdependence of living things: how vital and yet how fragile these relationships are.

65

Seaweed is an unlikely symbol of our new-found ecological aware-ness, but in fact it plays an essential role in our survival. For one thing, it helps to regulate our climate. Kelp likes to stay damp, and on a cloudy, overcast day it can remain 'comfortable', even at low tide. But on a sunny day, when it's 'stressed' and in danger of drying out, it releases iodine, a powerful antioxidant, from stores inside its tissues. This protects the kelp from harm, and it also has a secondary effect: the iodine rises into the atmosphere and contrib-utes to the formation of clouds. In rocky parts of the coast, where there are extensive kelp beds, such as Anglesey and the west coast of Scotland, seaweed is thought to be responsible for thicker, longer-lasting cloud cover – the stereotypical British seaside weather. How apt that the word *wrack* has another, less familiar use, meaning 'a mass of cloud'.

If in the course of opening our eyes to environmental realities we have lost some of our simple pleasures and amazements, we have replaced them with passionate collective attachments to a few powerful symbols of what is precious and threatened: the polar bear, the tiger, the wildflower meadow, and so on. Our lenses are not so often focused on the mucous, the smelly and the common-place. 'There is nothing more vile than seaweed,' claimed Virgil, and plenty of people since have felt the same way. Yet between them the humble algae produce more oxygen than all the plants in the world put together, making them crucial to life on earth.

We're still learning. No doubt algae hold other strange secrets we haven't even guessed at yet. Meanwhile, underwater photog-raphy reveals beauty, variety and incredible abundance: vast forests of seaweed pitching and threshing in the currents like birches in the wind, and teeming with life just as wild and sundry as the forests we know on land.

Images like this seem a world away from the odds and ends of debris I foraged for this morning. Is it any wonder we don't recognise it as the precious, the enigmatic, the true stuff of life, when it turns up on our beaches slimy, reeking and common as muck?

Summer

Denatured

Sand is becoming my element. Whatever else I find during my walks on the beach, it's the one thing I always bring home. I bring it on my boots, in the folds of my coat, the seams of my jeans. When I come in on a windy day, my hair and eyebrows are rimed with sand, and I can feel it between my teeth. The pages of my notebook are scratchy with it.

In a way, this is a return to childhood. When I think back to those childhood trips to the seaside, sand – with all its gritty ambiguity – is part of the very texture of those times. It's the exciting stuff of which our holidays were made, and it had several distinct manifestations. On the first morning of the holiday, there we would be, in Barmouth, Nefyn or Dale, making our way down a cliff path to the beach, laden with swimming things and towels and picnic and buckets and spades. The leader of our group had the task of identifying the prime spot and setting up camp, signalling to the rest by thrusting a spade into the sand like a flag at the North Pole. There were unspoken rules about this. We never, for instance, settled on the warm dry sand at the top of the beach, where there were crackly bits of sunburnt seaweed, and sandhoppers. It smelt a bit funny up there, and the sand was already used; you kept digging up someone else's old cigarette ends and lolly-sticks. No, the ideal space was flat and smooth and washed by the tide, where the sand was clean and firm for digging, and the hole you dug seeped cool water perfect for a castle moat.

Back then, sand was exotic. The only time we saw it at home was when we came back at the end of our week in Wales and emptied our beach shoes on the back step, making a miniature sandhill which remained a cherished souvenir of that other world, until the rain washed it away and summer was over.

But living in a place like this, you get used to living with sand. The wind drives it inland, where it scours the paint off window frames and parked cars. Upend the laundry basket, tip out all the socks and T-shirts onto the kitchen floor; what remains is sand. Empty the the vacuum cleaner and it's there, mixed with the fluff and dust which is the detritus of our lives. Even then I know there's plenty more of it, invisibly embedded deep in the fibres of every carpet in the house. When I clear the place out and move away, this is what will be left behind: a forgotten light bulb burning itself out in the cupboard under the stairs; the odd coin or hairgrip wedged between floorboards; and sand.

★

I'm watching the tide recede, and the deep channels emptying themselves, flowing back towards the sea. They're like rivers, waist deep in places, cutting through the sand, melting the edges soft until they collapse. Then they flatten out, spreading and dividing. The sea drags them under and into itself, clawing them back, garnering and collecting its property. It floats the razor-shells and wormcasts, the coal chips and seaweed fragments – all the dry stuff which has spent the morning lying in the sun, ticking and desiccating. It leaves the sand brushed and gleaming.

Five miles north of here, there was until recently a sand-winning plant. It closed down a few years ago, after a forty-year history of extracting sand from an area known as Horse Bank. The site of the old plant is now being reclaimed, and efforts are under way to reinstate at least part of it to the level of the surrounding saltmarsh, to allow natural vegetation to colonise it once more. The bird life is exceptionally rich here, and the RSPB has built hides for the birdwatchers who gather at weekends. The character of the place is undergoing a rapid transformation, from scruffy industrial blot on the landscape to nature reserve.

Back then, I assumed that the expression 'sand-winning' had a particular technical meaning – something a bit like winnowing, perhaps. I vaguely thought it might be to do with sifting or sorting. But in fact it is a simple use of the familiar verb. To win sand is to gain it, to work it out of the ground and take it away.

There's something about this word which hints at the hard physical labour involved; it's suggestive of competition, even of battle. Until recent times, the job was done by hand, the sand dug out with a shovel and loaded onto a cart. Sand is heavy, difficult stuff to remove and transport; to the men doing the shovelling and loading, it was hard won. More generally, it describes one aspect of the relationship human beings have had, throughout history, with the coastal landscape. The sea has been an adversary, whether carrying away objects and even people from the safety of dry land, or invading towns and homes. We have built defences, but the sea has breached them. We have learnt ever more sophisticated ways of mitigating the effects of its destructive power, but we know we can never tame it.

In the early days, sand-winning here was not well regulated, and environmental damage was done. Deep pits and channels were dug out between the sand dunes, in the mistaken belief that these would

73

fill in quickly without any long-term effects. Later, the fragility of this ecosystem became better understood and more highly valued, but there were still environmental implications to the work, including the toing and froing of heavy lorries through sensitive areas. It's a mark of how lax attitudes were in the past that no one knows for sure how much sand has been removed altogether, but during the more regulated history of the plant permission was granted for the removal of up to four hundred thousand tons a year. The earliest demand came from the construction trade, and huge quantities went into the building of new houses in Liverpool, but local sand has exceptional qualities for use in foundries and, especially, for polishing glass.

Like every child who has ever played on a beach, I know that sand is not one thing but many. Grains of mica would glitter on the surface, where I wrote my name with a finger. My plastic spade would cut down through alternating layers of darker and lighter sand, with the crispness of soft brown sugar or the thickness of molasses. Sand is a rich mixture of fragments, worn down from a variety of rocks and shells and the dry remains of sea creatures; it varies greatly from one coast to another, according to local geology. The sand on this beach is unusually high in silica, which is why it was extracted for use in glassmaking; silica sand is the major ingredient of all types of glass products, from tableware to mirrors and windscreens, from loft insulation material to plasma screens.

The extraction of sand from this beach has long been a source of controversy, and in 1998 it briefly hit the headlines in the national press. The occasion of this moment of notoriety was caused by remarks by the local MP, who spoke out against the council for allowing the practice to continue. He suggested that the sand-winning might be connected to the growth of 'mud and grasses' on the shore, but a council spokesman described this claim as 'rubbish', citing the fact that sand was accreting from the south, and going on to claim that 'the whole beach area has risen six feet over 40 years, which is a huge amount of sand'.

Journalists, naturally, were less interested in this kind of detail, and more interested in a different angle: the export of Southport sand to Saudi Arabia (*Sea resort sells sand to Arabs*, as one headline put it). I'd always thought this 'coals to Newcastle' story was a myth, but it turned out to be true. Saudi Arabians have plenty of sand, of course, but it doesn't have the right physical and chemical properties for all the specialist uses they need it for. As Michael Welland says in his surprisingly fascinating book *Sand: The Never-Ending Story*, 'international trade in sand would hardly seem to constitute significant commerce – after all, everyone has sand – but it does, because often the right kind of sand is in the wrong place'.

*

As well as taking sand away, human intervention can add something to it.

Walking on this beach, your attention elsewhere – watching a flock of oystercatchers, perhaps, or screwing up your eyes against the sun to see the hazy form of a container ship emerging from the Mersey estuary in the distance – you might stumble on a boulder. Nothing unusual in that, you'd think. But if you know the area you'll be aware that there are no stones to be found on this stretch of coast; it's entirely composed of soft, granular deposits of sand, silt, clay and peat. There are no outcrops of rock on this shoreline at all. So where has this boulder come from? And that one? Now you see there are dozens of them, scattered on the sand towards the top of the beach. They all look very much alike. You turn one over with your boot, give it an experimental kick, and something peculiar happens: it breaks. Funny sort of rock. You bend down and touch it. It's soft inside and a rich brown colour, crumbly in texture, and fibrous. It has a faintly warm and familiar scent.

Turning your back on the sea, and looking at the face of the sand dunes, you'll notice dark brown veins running horizontally through them, jutting out a little from the sand like rocky outcrops. These are the locally infamous 'Formby Cliffs'. But appearances are deceptive. They're not conventional cliffs, and the rubble of boulders on the sand beneath is not rocks. It's tobacco.

Some beach finds have made long journeys to get here; the tobacco boulders have travelled just a few metres, tumbling out of the front of the dunes onto the sand. They're part of an enormous tip, where huge quantities of tobacco waste were dumped in the dunes between 1956 and 1974. As the sea encroaches and the dunes erode, the waste is becoming exposed and spilling out.

Its recent travels were modest, but the story of this material places it amongst the most exotic of all the things I've ever found here. It's impossible to know where it originated; the tobacco plant has so many varieties, and is so versatile, that it's grown all over the world, from China to Brazil, from Cuba to Poland, from India to the USA. Its presence here has everything to do with the tremendous success of the Port of Liverpool just down the coast. Liverpool has had a long association with the tobacco industry; as well as sugar, cloth and grain, tobacco was one of the principal imports, and the tobacco warehouse on Stanley Dock, which now stands empty, was said to be the largest building in the world when it was completed in 1901. A stone's throw from Pier Head, where Formby asparagus was being loaded onto the luxury transatlantic liners, hogshead barrels of tobacco were being unloaded onto the docks.

It made sense to set up processing and manufacture nearby. A family firm called Ogden's started up in Bootle, introducing its most famous brand, St Bruno pipe tobacco, in 1896. By the turn of the century, the factory was producing 900 million cigarettes a year. Ogden's was right at the cutting edge of the industry; it even pioneered the introduction of collectable cigarette cards or 'stiffeners', and for a time its golden ticket of modern production methods and highly successful marketing made it look unstoppable. Then in 1901 an American manufacturer launched an aggressive takeover bid for the entire UK industry, sparking what became known as the Tobacco War. The only way the British companies could survive was to band together, and out of that unpalatable but pragmatic decision the conglomerate Imperial Tobacco was born. The names of some of its constituent companies, such as John Player & Sons and Lambert & Butler, live on today as famous brands.

Manufacture creates waste, and the burgeoning Imperial Tobacco had a major problem on its hands in the shape of enormous quantities of bulky material which had to be disposed of. Tobacco companies go to some lengths to minimise the waste of their expensive raw material. At the start of manufacturing, tobacco leaf and stem are separated and go through a process of mixing, wetting, blending and cutting. Along with a substance called 'reconstituted tobacco sheet', they are then further mixed, dried and cooled to produce 'rag', the stuff which goes into cigarettes. Spilt rag from damaged cigarettes is collected and reprocessed, but nevertheless there are large quantities of 'offal', made up of dust and waste tobacco. Some of this is salvaged and made into reconstituted tobacco sheet; the rest is dumped.

In some parts of the world, tobacco dumping is a real-life horror story. The global tobacco industry produces over two billion kilograms of manufacturing waste and two hundred million kilograms of chemical waste every year. As ever, the worst consequences of waste disposal are borne by the developing world, where resources are limited and regulations less strictly enforced. In Malawi, where a third of the premium tobacco consumed by British smokers is produced, children as young as five years old pick through mountains of waste tobacco, sweeping up the scraps with their bare hands in the hope of selling them and scraping a living. They work in a perpetual haze of dust, which causes breathing difficulties and problems with their eyes and skin. The true nicotine levels in the waste are often unreported, so it's hard to know just how toxic it is.

The stuff dumped here at Formby is sometimes mistakenly described as 'nicotine waste'. In fact it had been 'denatured', so that the nicotine content was minimal. Nevertheless, there were other environmental effects, and the re-emergence of this material, thirty-six years after the last load was dumped, has brought back to public attention a highly controversial chapter in the history of the management of this coast.

*

In their book *Edgelands* (2011), Paul Farley and Michael Symmons Roberts sing the praises of forgotten and overlooked spaces outside our towns and cities, spaces which are neither urban nor rural. I could make a case for the beach as a kind of edgeland. It's a literal edge, of course. It's liminal, but not unspoilt. There's litter, there's joyriding; it's still possible to see the historic evidence of what little industrial activity has been possible here – sand extraction, tipping. But it's essentially intractable land, unreliable and unproductive, so its heart remains wild.

Like the sand-winning, the dumping went on for years without planning permission. The waste was tipped onto an area which had previously been used for asparagus beds. Formby had been a famous centre of asparagus growing, but in recent years production had begun to decline, because the advancing sea was gradually spoiling the fields by making the land too saline. Presumably, then, the asparagus beds were seen as no great loss. They were falling out of cultivation, and looked like an available and convenient site for tipping. And so it was that in 1956 the British Nicotine Company, a division of Imperial Tobacco, began to drive trucks into the fields and dump 22,000 wet tons of tobacco waste per year onto a seven-hectare site. It was spread in layers and left to dry, then bulldozed to the perimeter of the site and mixed with sand.

In those years while dumping proceeded without permission, it was nevertheless tolerated, even welcomed, by local people. There were dissenting voices, but they were drowned out. This is partly down to the different attitude to environmental protection at that time; it's also testament to the success of the public relations campaign mounted by Imperial Tobacco, which made its waste sound like the answer to a problem. Promises were made, and they were believed, not least because there had been other interventions in the past which had indeed brought benefits of one kind or another here in this difficult coastal territory.

In 1964 the argument was made that tobacco waste would help make the dunes more resistant to coastal erosion. The dunes act as a sea defence, protecting agricultural land and homes, but they are by their very nature fragile and shifting. The tobacco was a bulky, solid material, and it was said that if it was mixed with sand it would change its texture, making it less mobile, less capricious and less susceptible to the sea.

This was a powerful argument; there have always been anxieties about erosion and what it might mean for communities living in the shelter of the dunes. Local people had reason to be acutely aware of the dangers; from the turn of the century the coastline had been retreating rapidly, and in the last few years a caravan park and a café had both been lost to the sea. Meanwhile, another ambitious project, the planting of pine forest on the rearward dunes in the late nineteenth century, had transformed the character of the area but had succeeded to some extent in stabilising the land. No wonder people were receptive to a new development that might make a difference.

The reality has not in any way lived up to Imperial Tobacco's promise. Monitoring and research has shown that the dunes in this area are no less susceptible to the erosive power of the sea; the tobacco cliffs are being pushed back at exactly the same rate as the natural dunes. Much worse, the cliffs present a serious unforeseen problem: far from improving sea defence, they hamper it. In normal circumstances, windblown sand builds up further inland, meaning that the dunes are not destroyed but simply move back as the coastline retreats. This is the natural pattern. But the compacted tobacco waste behaves differently; it presents a more solid and cohesive feature which can't be blown back freely by the wind, can't be 'stored in the system'. It actually prevents rollback.

Another of Imperial Tobacco's claims at the time was that the waste, non-toxic and rich in humus, would improve soil conditions and attract wildlife. Again, there was a precedent for the use of soil additives in the area. Indeed, the entire asparagus boom was built on a very particular sort of fertiliser: human excrement. When the railway first arrived here in the 1840s, a local businessman, Thomas

Fresh, started bringing large quantities of 'night soil' from Liverpool to be used as a fertiliser. Liverpool's sewerage system was still being constructed, so there was no shortage of the stuff. The sudden availability of this regular and plentiful supply gave local farmers the chance to bring areas of the rearward dunes into cultivation for the first time.

The business of collecting, transporting and spreading human faeces may not seem the answer to your prayers, or mine; but these people were neither proud nor prosperous enough to turn down the opportunity. The sandy soil had so far been too poor to grow any crop; there had been some rabbit farming, but otherwise this land was lying idle. Now there was a chance to bring it into a condition where it could yield a desirable crop and a living for those families which took up the challenge. The enterprise was brilliantly successful. At its height the asparagus crop accounted for two hundred acres and was an important regional delicacy, winning awards year after year and finding its way onto the plates of wealthy travellers on the transatlantic liners.

It's easy to see why, with the sea encroaching on the dunes and the asparagus fields lying derelict, the tobacco waste might have seemed like a solution. It would condition the soil, bringing structure and nutrients. Perhaps there was an opportunity here to revive the productivity of this land, if not to return to the glory days.

★

You're never alone with a Strand, ran the advertising slogan for one of Imperial Tobacco's brands. A Frank Sinatra lookalike wandered moodily through the streets of London, pausing under a street lamp to light a cigarette. The campaign backfired spectacularly: smokers wanted nothing to do with a brand that marked them out as loners, and Strand cigarettes were withdrawn from sale a few years later.

Nicotine Path and Nicotine Wood are on the map now – not, as visitors might think, smoking areas, but the legacy of twenty years of tipping. Nicotine Path leads from the National Trust car park through the wood and then the dunes and out onto the shore. Walking here in August, I'm astonished by the stark visual contrast between this area, where the soil has been changed, and the natural duneland which surrounds it. From a high vantage point, I can see a sharp, clear dividing line, as straight as if it had been drawn using a ruler. On one side of the line, the sparse grey-green of marram grass on sand; on the other, hard up against it, a bright emerald-green ocean of stinging nettles. This vivid colour looks lush and rich, but the area is impoverished in terms of indigenous wildlife. It's dense with vegetation, but the plants which thrive here are alien species: nettles, thistles, bur chervil and a handful of other agricultural weeds that are completely absent from the surrounding duneland.

The incomers have displaced native species of flora, and driven out the fauna, except for rabbits, which seem just as able to proliferate here as elsewhere. Now, at the height of summer, the ground is all dry, but in winter there are pools here, stained brown by tobacco. The brown discoloration deprives the water of light and oxygen, which makes it unsuitable as a habitat for most of the creatures that populate the dune slacks.

Back in 1964, there were a few isolated voices raising concerns, based on the value of the area as a recreational amenity for local people and visitors. But the giant won the day. One of the conditions was that the site must be landscaped to mimic the natural shape of the dunes. So there are some ups and downs, not completely unconvincing. But the resemblance stops there. If I dig a little into the ground with my boot, I immediately expose darker layers, huge clumps of tobacco interleaved with the sand. Whatever methods were used to mix the two materials, they were not effective.

This place has seen so much human drama. Success and catastrophe. Hope and cynicism. Determination and despair. I'm standing right on the site of the old asparagus beds, the location where the legendary crop was raised, a piece of land now so perilously close to the sea that it's crumbling and falling onto the shore. This is the place where salt water gradually destroyed the fertility of the soil, so carefully built and nurtured from the chamber pots of Liverpool.

By all measures, tobacco tipping was bad for the environment. Why wasn't it stopped? I raised this with Andrew Brockbank from the National Trust at Formby Point. He has some great stories to tell about the shifting dunes and what's hidden inside them. Brick walls; bits of old car park; even a whole caravan, which worked its way to the front like a splinter working its way to the surface of the skin, dangled there for a day or two and then fell out onto the beach.

'Yes,' he said, 'but who talked about the environment back then? They called them "cranks".'

He's right – that was the word. It's one you don't hear so much these days. The early environmentalists were well aware of how the rest of society saw them, but they were not discouraged. Ernst Schumacher, author of Small is Beautiful, famously said that he didn't mind being called a crank 'because a crank is a small, simple and efficient tool that makes revolutions'.

But that was in 1973. The revolution was a long way from this neck of the woods in 1960, when Imperial Tobacco applied, somewhat belatedly, for planning permission to go on tipping. Permission was granted. The agreement was terminated four years later, and an inquiry was held to determine whether the practice should continue. The inquiry too found in favour. It wasn't until 1974 that the lorries stopped coming, and by then 400,000 tons of tobacco had been dumped in the middle of what is now a nature reserve. Then it was quiet again in the dunes, and nettles grew and covered up the mess.

But they hadn't reckoned with the sand: its shiftiness, its talent for surprise. Sand and sea have this in common: they will swallow our dirty secrets, but they cough them up again in the end. Whatever it is, we may think we've chucked it, buried it, seen the last. Decades later, the whole unfortunate business is forgotten or glossed over. And now suddenly, here it is, waiting to be found, lying in plain view: the evidence.

Poor Man's Asparagus

The saltmarshes and mudflats just up the coast have long been known for their samphire, but this is the first time I've ever seen it cast so far south. A high tide has torn it up by the roots, dragged it a few miles and flung it up onto the sand, where it lies in small tangled heaps. Some are bright and verdant, others already a bit bedraggled and visited by flies. I gather a few of the freshest heads to take home, but when it comes to cooking it I've changed my mind. Samphire wilts and deteriorates quickly; it should be cut from the living plant.

The word 'samphire' is a lovely one – evocative of jewels and fire and with a vaguely Elizabethan ring to it. It's said to be a corruption of 'St Pierre', in tribute to the patron saint of fishermen. It's known locally by the somewhat less poetic 'Sampi' or 'Sand Pea'. But it's almost exclusively the older residents who know these traditional names, and who remember what it tastes like. The practice of collecting your own food from this shore has been obsolete for at least a generation.

Now it's beginning to make a comeback, though it's slow to catch on among local people. There are stories in the press about 'families of Indians' visiting the town, gathering on the beach to collect samphire, 'which they use as food for their families'. This, promises the caption to one picture, is likely to become 'a regular sight' for locals.

A few enlightened restaurants in the area have even begun to put it on the menu; it goes especially well with the local shrimps, apparently. Insofar as this might amount to a revival in the fortunes of local samphire, it's typical: as with laverbread, so with samphire. An ordinary wild food – something you can forage for, free of charge – falls almost completely into obscurity, and is then reinvented as an expensive menu item.

But in this case there's another, more compelling reason why it skipped a generation. From 1972 until 1996, a herbicide called Dalapon was applied to areas of the shore. The effect of spraying was to control and eventually eradicate all spartina grass and samphire from the affected area. It continued to grow on unsprayed areas further north, but it disappeared entirely from Southport itself.

Dalapon has a sinister history; it's best known as a major ingredient in the notorious Agent Orange, a defoliant sprayed from American helicopters to strip forest and agricultural land during the Vietnam War. On its own, however, it has a reputation as a relatively safe chemical, and these days it's used to kill grass weeds, principally in sugar cultivation.

The use of Dalapon here was a response to concerns that the saltmarsh was growing and spreading over sandy areas of beach, with a loss of 'amenity value' for local people and holidaymakers. It was just one episode in a long-running conflict over management of the saltmarsh, which continues to be controversial to this day.

Now, over a decade since spraying stopped and the marsh was once again allowed to define its own borders, the samphire is back. It carpets the shore in vivid emerald from June to August, and the people who live here are starting to relearn its value by listening to the reminiscences of their parents and grandparents.

*

The use of samphire as food goes back a long way. In *King Lear*, Edgar peers down a dizzying cliff at Dover and spots a figure:

> half-way down
> Hangs one that gathers samphire, dreadful trade!

The species he's referring to is rock samphire or *Crithmum maritimum*, which grows less abundantly and in less accessible spots. The man clinging to the cliff face may well have been risking life and limb to gather it, which suggests that it was very highly prized. It tastes strong and rather bitter: 'a mixture of celery and kerosene', according to one authority on the subject.

Marsh samphire or *Salicornia* is more subtle in flavour, and harvesting it is a less hazardous activity, though a very muddy one. The seventeenth-century herbalist Culpeper wrote approvingly of the taste, and claimed it was useful in curing 'ill digestions and obstructions'. In the past it was pickled and eaten by sailors on long voyages to help combat scurvy. It was popular with city-dwellers as well as coastal people; as long ago as the nineteenth century, it was shipped in casks of seawater from the Isle of Wight to market in London. It used to be known as 'poor man's asparagus', but these days it frequently fetches twice the price in London shops. I don't think it travels well, either – the crispness is quickly lost and you can be left with a flabby mouthful of salt. Much better to reserve it for a treat when you're visiting the seaside: go out with a pair of scissors and harvest it yourself, take it home immediately, steam it for the barest couple of minutes and serve it with melted butter.

Poor man's asparagus has one considerable advantage over the rich man's version, but only for half of the population. About 50 per cent of people metabolise asparagus in such a way that it makes their urine smell strange. The rest of us don't know what the 50 per cent is complaining about, since we lack the digestive enzyme that breaks down a compound called mercaptan, responsible for the phenomenon.

Bizarrely, this smell is also known for its ability to attract fish. I'm not sure how this predilection was ever discovered, but according to legend it has been put to an interesting practical use. It's said that during the Second World War American pilots were given tinned asparagus in their rations, in case they found themselves shot down and stranded on remote islands. All they then had to do was pee in the water; fish would be irresistibly drawn towards them, and into their improvised nets. A Flying Fortress bomber had a crew of ten; given the genetic odds, if only half of them survived the crash landing, you'd have to hope it was the right half.

Incidentally, mercaptan is the same compound that gives skunks their notorious stink, a piece of information which casts a doubtful light on the reputation of asparagus as an aphrodisiac. But samphire has no such side effect. Perhaps dinner party hosts should serve both vegetables as alternatives, passing the two dishes and raising an eyebrow delicately to indicate which is for *those who can* and which for *those who can't*?

Aphrodite

'Who's going to pick it up?'

Neither of us has any idea what we're looking at. It's an object about four inches long, oval in shape, completely covered in a felted mass of russet-brown fur, and fringed all around with longer, silky hairs in shades of iridescent green and bronze. It's exquisitely beautiful, like a strange piece of vintage jewellery . . . and yet at the same time it is slightly alarming, even repellent. Perhaps it's that iridescence: it's so unexpected, so unnatural-looking. It gives it a weirdly high-tech appearance. Of course I know this must be a creature of some kind, but it looks so alien that I can't make any sense of it at first.

We've been crouching on the sand and staring in silence for ages, but curiosity is winning the day. Somehow, though, I'm reluctant to touch the mysterious object with my bare hands, and I can see my friend feels the same. Well, you never know – there are plenty of biting and stinging things about, after all. I find a lollystick, edge it underneath very gently and pause a moment to see whether there's any protest. The creature – whatever it is – remains stiff and impassive. It's clearly dead.

My walking companion is about to head back home to Sydney. Before she leaves, she's visiting friends in different parts of Britain, making the most of a couple of weeks' holiday, seeing as much of the country as she can. Our comfortable, freewheeling conversation was interrupted when we reached a dune-top vantage point, stopped and looked at the view. It's a glorious day, high summer. Two small fishing boats dawdled hazily on the still water; the Lennox oil rig eight miles offshore shimmered like a distant palace. We sat on the warm sand and watched dozens of brightly coloured power-kites arcing above the beach. One day I'm going to have a go at that, I said. From up there, it looked like the very definition of exuberance. When you get closer, the kiting is more combative: it's all muscular blokes, working with and against the kites, wrestling them into the wind, holding them there, paying out just enough and then pulling back hard, slamming them down onto the sand where they lie flapping and billowing, still rebellious but with the breath knocked out of them at last.

We sat watching for half an hour, then headed down from the dunes, the spiky marram grass stinging our bare legs. We stopped to speak to a man, unseasonably dressed in a green duffel coat and flying a remote-controlled plane – a *Richthofen's Fokker*, he told us earnestly, one twelfth, all balsa. We exchanged that broad vista – the sweep of the bay, the big sky – for the small details of the strandline, pausing to look at the assortment of stuff the tide had brought in, marvelling again at the sheer variety of it. We found a bicycle saddle, a knitting needle, a large bleached knuckle bone, a light bulb . . . and our mystery creature.

When I turn it over with the lolly-stick, it's obvious that this is not some small mammal, as I first guessed. Its underside is hairless and fleshy, a dull yellowish colour. From this perspective it's not unlike a very large slug, but more textured, ridged like the surface of a worn tyre.

I'm in the habit of carrying a plastic bag on these walks, for just this sort of discovery. I push the tiny creature in. It's suddenly dull and uninteresting in the bottom of the bag, like a wad of damp sand. I wrap it up and tuck it away in the rucksack to be taken home and inspected.

Back home, I put the kettle on and my friend flicks through reference books, exclaiming over the illustrations. She had no idea barnacles were so beautiful! And could I guess how many different species of crab there are on the British coast? Then she gives a shout of triumph: our mystery is solved. We've found a sea mouse.

The sea mouse is a scaleworm. Its back is covered with over-lapping scales, completely hidden by the dense hair which covers them. It has two curved antennae or palps. It lives in clean or muddy sand in shallow water, burrowing just beneath the surface and ploughing along, hoovering up its rather indiscriminate diet of carrion, detritus, microscopic animals and other, smaller types of worm. This basic information is all we have to go on for now, and meanwhile our specimen is starting to stink. It has that rich, intense smell of everything that comes from the sea, but this animal is well and truly dead, and there's a whiff of decay which is putting us off our tea. Once we've photographed it for posterity, I dig a small hole in the garden and bury it, a wild and strange little corpse amongst the venerable remains of pet gerbils and rabbits and goldfish.

*

You can keep a peacock feather for years, and the blue will not fade; there are butterfly samples with the colours still as bright as the day they were caught and pinned out in display cases by Victorian collectors. The reason for the longevity of these colours is that they are a sort of optical illusion, made up not of pigment but of photo-tonic crystal structures generated within scales on the wing or the feather. The same phenomenon is responsible for the sea mouse's fantastic skirt of rainbow-coloured hairs. Each hair contains a microstructure made up of hexagonal cells. The crystal-like structure is incredibly efficient in trapping light photons, bouncing them around the particles within the cells and causing them to emit unusual colour frequencies.

In the case of the sea mouse, no one is certain what this 'structural colour' is for. In one book I read that the purpose is probably to send a warning signal to predators convincing them that the 'spines' (which are actually harmless hairs) are poisonous.

No sooner have I accepted this plausible explanation than I come across an article in the *Journal of Clinical Pathology* entitled 'Severe granulomatous arthritis due to spinous injury by a "sea mouse" annelid worm'. This is a coldly scientific but nevertheless gruesome account of an injury sustained by a young man snorkelling in the Mediterranean. His right index finger was so swollen and painfully damaged that after eight months it was amputated, and there was found to be 'exuberant granulomatous and foreign body type inflammation in the dermis and subcutaneous tissues and affecting the bone, with erosion of the cartilaginous surfaces of the proximal interphalangeal joint'. Laboratory analysis showed traces of the fibrous substance chitin embedded in the amputated finger, and it was concluded that these were 'almost certainly' fragments of fibres from a sea mouse, rather than a sea urchin as originally suspected.

To the inexpert eye, it's hard to imagine that a sea mouse could really be responsible; the hairs on its body seem quite fine and silky rather than spiny. If it's true, it could knock a hole in the warning signal theory. On the other hand, colour in the animal world is usually explained in terms of display, but since this animal spends its time under the surface of the seabed it must have limited opportunities for showing off. No one has the answers yet.

The taxonomist Linnaeus was familiar with this species; he christened it *Aphrodita aculeata*. Aphrodite, the Greek goddess of love, takes her name from the Greek for 'sea foam', and it would be nice to think that borrowing this name and applying it to the sea mouse was a tribute at once to the creature's peculiar beauty and to its oceanic origins. Disappointingly, it turns out to be a rather feeble dirty joke, relying on its alleged resemblance to female genitals. I'd say that it requires a fairly spectacular leap of imagination to see the similarity. Perhaps the great man of science was less of an expert in such matters than he liked to think?

You could say Linnaeus had a reputation to live up to. He had already made himself notorious by insisting that sexual reproduction took place between plants, scandalising the eighteenth-century Church and even some of his fellow scientists. Far from tiptoeing around this controversial subject, he wrote of it in the most provocative terms:

Love comes even to the plants. Males and females hold their nuptials; showing by their sexual organs which are males, which females. The flowers' leaves serve as a bridal bed, which the Creator has so gloriously arranged, adorned with such noble bed curtains, and perfumed with so many soft scents that the bridegroom with his bride might there celebrate their nuptials with so much the greater solemnity.

He even went so far as to name a genus of plants *Clitoria*. He was accused of 'loathsome harlotry', and William Goodenough, one of his contemporaries, declared himself appalled by the man's 'disgusting names, his nomenclatural wantonness, vulgar lascivi-ousness, and the gross prurience of his mind'.

Plant reproduction lost its shock value a long time ago, and Linnaeus's rather purple prose is unlikely to offend anyone these days. But the American natural history writer Sue Hubbell describes him as 'an unpleasant man' who 'often made mean jokes in his namings'. In her book *Waiting for Aphrodite: Journeys into the Time before Bones*, she turns the tables on this meanness by recounting a lesser-known variant of the Aphrodite myth. In this version, the goddess takes on masculine traits, such as a beard and male sexual organs, and is worshipped fervently in ancient Cyprus. So the sea mouse has as part of its nomenclature 'the androgynous qualities of an undifferentiated creative principle, of bountifulness, made concrete as a Great Goddess with mascu-line traits . . . not bad for a six-inch fuzzy iridescent worm of ancient lineage from the ocean deeps'.

Hubbell's book describes her long preoccupation with the sea mouse, and her fervent wish to see a live specimen for herself. This quest brings her into conversation with fishermen and naturalists, but what she learns is that surprisingly little is known about this creature. It's quite a challenge, studying a species which lives in the mud on the deep ocean floor. When it does make an appearance on land, it's stranded and generally dead or dying. As one specialist scientist remarks to Hubbell: 'No one has ever seen Aphrodite except when it was unhappy'. Perhaps it's not surprising, then, that its reproductive life is especially poorly understood. No one knows how it broods and hatches. No one has seen a newly hatched specimen.

At least, no one knew in 1999, when *Waiting for Aphrodite* was published. It's possible that things are different now. In the decade since, a bright spotlight has been shining on the sea mouse, drawing a glint of new promise from those iridescent hairs. My friend, now back home in Sydney, sends me a newspaper clipping about a recent study which has revealed the sea mouse's potential to help in the redesign of fibre optic technology. Scientists in Australia, examining the fibres on its body, recognised a honeycomb pattern of holes in the cell structure which were even more effective than the ones being made in the laboratory in work to develop new and better types of optical fibre. They began to investigate how they could learn from nature. 'These photonic crystal fibres could be the answer to the world's increasingly heavy demand on communications,' according to the report. 'Fibre optics have been threaded throughout the world for years to enable information to travel more freely. The new fibres could carry much more at a fraction of the cost.'

And so the sea mouse, by nature a private and unobtrusive creature, is thrust suddenly into the limelight. It's captured and studied and experimented upon, and it won't be long before its mysteries are all known. Still, it's the very strangeness and secretiveness of this animal which captivated Sue Hubbell.

Strangeness is the key, too, for the American poet Amy Clampitt. Her poem 'Sea Mouse' opens with the same sense of the new and the rare that I experienced when I first saw this creature:

> The orphanage of possibility
> has had to be expanded to
> admit the sea mouse.

That rich phrase 'the orphanage of possibility' imagines a solitary individual, without the usual context of family or ancestry. It's as if the one in the poem, found 'sheltering under ruching / edges of sea lettuce', might be the only specimen in existence: a biological one-off.

I felt something similar as we turned ours over with the lolly-stick; I'd never seen anything like it. The oddity of this find seemed at first to put it outside any of the familiar categories of living things – not fish, not amphibian, not mammal, in spite of its misleading nickname – and to make it anomalous, even unique. In Clampitt's poem, it is a 'foundling', a swaddled and 'pettable' infant. The one we found had been dead a little too long to be considered pettable, but still I know what she means: its smallness and roundedness, that furry pelt.

'Sea Mouse' is one of a number of poems Clampitt wrote while staying on the Atlantic coast in Maine in the 1980s. At the heart of each poem is something observed on this shore which she had come to know so well. Her poem 'Beach Glass' celebrates equally the fragments she finds, however humble their origins: 'amber of Budweiser' and 'lapis' of the Milk of Magnesia bottle, turned by the sea:

> as gravely
> and gradually as an intellect
> engaged in the hazardous
> redefinition of structures
> no one has yet looked at.

The double meaning of the word 'redefinition' is beautiful: the sea is literally remaking the outline or shape of the glass by smoothing it, and the poet is engaged in parallel activities: the careful handling and examination of small, fragmentary and apparently unremarkable things, and the development of new ways of describing or understanding them.

*

It's sobering to realise that one of our most complex and impressive human achievements – the development of high-tech communications technology – is outdone by an obscure mud-dwelling worm, a creature which shares the ocean floor with the fibre optic cables we've laid there.

Those same cables were brought ashore here, on this very beach. For ten years, this has been the UK's cable landing station for a transatlantic submarine communications system linking Canada, the USA, Ireland and Europe; the latest phase of cable laying, known as Project Kelvin, is now close to completion. The other landing stations are at Herring Cove in Nova Scotia, Lynn in Massachusetts, Dublin and Coleraine. The link allows direct access to American telecom networks without the need for communications traffic to be transmitted down to London and routed across the Atlantic via Cornwall. The benefit to the north-west of England, in the language of Hibernia Atlantic, the company that owns the system, is that businesses in the region can now 'feasibly achieve greater productivity by building "follow the sun" shared working regimes, particularly for time-critical research and development projects'.

Project Kelvin takes its name from Lord Kelvin, the Belfast-born scientist William Thomson, a pioneer of electric telegraph engineering who advised on the laying of the first viable transatlantic cable in 1866. The original cable linked Valentia Island, off the south coast of Kerry, with Newfoundland, and was described at the time as the 'eighth wonder of the world'. It would be hard to exaggerate just how revolutionary it seemed, though it had been in the minds of scientists and inventors ever since the invention of the telegraph thirty years earlier. Some of them, including Samuel Morse, had been busy with their own experiments, submerging wires and attempting to send telegraph signals through them. The main problem in these early days was insulation: tarred hemp and rubber were both tried, but found to be insufficiently waterproof. It was the discovery of gutta-percha, an adhesive gum that could be melted and applied to wire, which provided the breakthrough.

The laying of the transatlantic cable was, not surprisingly, an event of enormous public interest and excitement. The *Illustrated London News* carried a detailed and colourful account of the Kerry end of the operation, which seems to have involved the efforts of just about every able-bodied man in Ireland. The last twenty-seven miles of cable was constructed separately in Woolwich and brought by steamer to Valentia, in such a great storm that the ship almost sank and had to be towed the rest of the way. The tow rope broke and the whole enterprise very nearly collapsed, but the cable was eventually brought close to shore. The attaching of the cable to the telegraph poles was then delayed when one of the poles was demolished by a cow (or possibly a drunk driver, depending on whose word we take for it).

Once the steamer had arrived, the shore section of the cable had to be landed on the wild and rugged island, passed hand to hand from ship to shore by men on a 'floating bridge' of boats. The first attempt failed because of the excessive enthusiasm of the men, who in a storm of *hurrahs* threw it overboard prematurely; the whole length of it had to be hauled back in and the process started again. 'Numbers of men were in the water up to their waists or shoulders easing the cable over the rocks,' writes our correspondent, 'while along the steep path up the cliffs was a close row of figures, men and boys, of every rank, from the well-to-do farmer down to the poorest cottier, all pulling at the cable with a will.' The cable had to be driven into a groove cut into the face of the cliff, and its massive coils carried across the meadows, roads, hedges and ditches to the Telegraph House.

The following day the battered steamer carried the last half-mile of the shore end of cable fifteen miles out to sea, where a much larger ship was waiting, carrying the main length of 2,300 miles. The cable was wound on board and the work of splicing the two ends together began. The outer layers were stripped away, exposing the copper conductor, which was then pared down to a fine wedge-shaped point at each end. These were joined by hand, bound with wire, soldered and sealed using multiple layers of gutta-percha, hemp and more wire. This laborious task took several hours, and then the cable had to be tested for conductivity before finally being cast adrift.

When the cables arrived here on my beach, there was no bridge of boats, no men standing up to their shoulders in water, no stripping and paring of the cable by hand. Modern fibre optic cables are laid by special purpose-built ships, fitted with on-board laboratories where the splicing and testing take place. The ships have extra propellers which make it possible to manoeuvre with great accuracy. This is crucial, since fibre optic cable, unlike its early copper antecedents, is not very malleable and has to be laid out dead straight from the stern.

The cable is designed to carry digital payloads both for telephone and internet traffic. It provides, in the jargon, 'a low-latency, self-healing, secure link'. There's something soft and cuddly about this language that makes it sound more like holding hands than light pulsing through fibres, but this technology is one of the building blocks of twenty-first-century life all over the world. Submarine fibre optic cables connect every continent except Antarctica, where the extreme cold presents as yet unsolved technical challenges.

Elsewhere, there are challenges too. The cables are not without their vulnerabilities. They don't entangle whales, as the old-style telegraph cables used to; but they have been deliberately cut as an act of sabotage in wartime, tapped for surveillance, broken in earthquakes, damaged by trawlers, bitten by sharks and stolen by pirates.

★

There are different ways of finding things. For Sue Hubbell, setting eyes on a sea mouse was the culmination of years of searching. She had heard of its existence, but since it's an undersea burrowing species with a secretive lifestyle she'd never seen it, and it had become to her mind almost a legendary creature. Old fishermen smiled mysteriously and said oh yes, they had seen them. They promised to catch one and bring it to her. The way they talked about them started to make them sound as fantastic as mermaids. 'No one forgets a sea mouse after he's seen it,' she declares.

There are things we go looking for, things we've heard of and longed to see for ourselves. But the best finds are often accidental. The discovery comes first, followed by the fascination, the need to know more. This characteristic of the beach – its capacity to surprise and mystify – is what brings me back here, day after day, month after month. I didn't even know such a thing existed until it was there on the sand in front of me and I was turning it over with a lolly-stick.

But Hubbell's right about the not forgetting. Nine months later,

I'm on the beach with my new husband after a March storm. The sand is littered with crabs and shellfish, jellyfish, starfish, more varieties of each than I can count: some dead, others injured and traumatised. They've been thrown ashore by the strong waves and stranded, desiccating, out of their element, for the gulls to snap up, prise open or tear to pieces.

We pick our way across this field of death, stepping carefully, pausing to watch razor-fish like pink tongues protruding from their long shells. The razor-shell must be one of the most common shells on British beaches, but they're more frequently found empty and abandoned. The tongue-like protrusion is a highly muscular foot, which the razor fish uses to dig itself into the sand; a healthy specimen can bury itself completely within a few seconds, but these damaged individuals were reduced to helpless squirming. It's a horribly mesmerising thing to watch. Then I spot a sea mouse.

This one is alive. It's lying on its back, moving almost imperceptibly. It reminds me, in a flash, of a tortoise which belonged to a neighbour of mine many years ago, and which would regularly tumble off the low dividing wall into our garden, where it would lie in the undergrowth, treading the air silently, in a gesture of uncomplaining despair.

It's no good turning the sea mouse over, as I did the tortoise. It needs to be in the water. Since this is a storm beach, and not a lolly-stick kind of day, I pick up a razor-shell (unoccupied) for a makeshift stretcher, and my husband nudges the casualty aboard, and heads, egg-and-spoon style, for the sea.

He's only gone a few yards when I stop him with a shout: I've found another! Also alive, also supine and helpless. He takes it aboard. Then a few yards further, another. And another. We find five sea mice in five minutes. The beach must be littered with them.

They lie on the sand, brown and insignificant like stones, but as we turn them and lift them they glint green and bronze, jewelled with light.

Each one, when we slide it into the shallow water, revives quickly and seems to feel the pull of home. It has endured its time under the glare of the sky, and wants only to return to obscurity. It begins to bury itself with a slow shuffling motion into the wet sand, until there's nothing left to see but a soft oval outline, disintegrating to smoothness under the in-and-out of the waves.

The Albatross and
the Toothbrush

In my scruffy old copy of *The Arrow Seaside Companion*, first published in 1956, Hugh Stoker lists some of the 'useful finds' he has made on his local beach in Dorset during twenty years of walks there:

> Several gold rings, over a hundredweight of lead fishing weights, revolver bullets and old musket balls, an ancient dagger and Roman coins (unearthed by cliff falls); about half a dozen ladders, a crate of canned Australian beer; a waterproof canister of tea; a football; a nylon parachute, a child's rocking-horse; a tinned chicken; a shove-ha'penny board; dozens of ship's brooms, dan buoys and lengths of rope – to mention just a few items which come to mind.

Even allowing for a little poetic licence, this list conjures up a fascinating world of lost things. The musket balls, dagger and coins are valued for their antiquity, but almost every other item also now has a whiff of the past about it. How many pubs still have shove-ha'penny boards? When was the last time a sailor swept a deck with a broom?

Stoker goes on to record a further count of 'rather startling things' found in the vicinity, including 'a German torpedo, a landmine, a capsized boat, the headless body of a man, two escaped Borstal boys, and a woman's leg, still clad in a silk stocking'.

The stocking is a delicious detail. But the striking thing about both these lists is the total absence of the one thing the contemporary reader should expect to find on every beach he sets foot on: plastic.

It's ubiquitous. Research carried out for the United Nations suggests that there are, on average, forty-six thousand pieces of plastic floating on or near the surface of every square mile of ocean in the world. And if it's in the sea, it's also on the shore. Every shore. Not only our own busy beaches with their populations of day trippers, but equally the most remote and undeveloped places, thousands of miles from where it started out. One such place – now something of a cause célèbre – is Midway. It may sound like an anonymous new town somewhere in the south of England, but Midway is a Pacific atoll, home to some of the world's most endangered species. It's a favourite haunt, for instance, of the albatross, a bird whose name is exceptionally rich in associations, in mythical as well as actual significance.

But recently, albatrosses have been dying here, many more of them than usual, and when researchers dissect the corpses and examine their stomach contents, what they find is a shocking variety of small plastic objects. Among the most commonly found are toothbrushes, cigarette lighters, Lego bricks and bottle tops.

It's the familiarity – the domesticity – of these small, disposable objects which breaks the heart.

*

Casual observation tells me that of the great variety of chocolate bars which have been eaten here on the beach in recent days, the runaway favourite is Cadbury's Time Out. Branding and advertising are more powerful influences than we like to think, slipping their messages into our minds like notes under closed doors. Holiday-makers queueing at the counter of the snack bar parked on the sand make their choices, and for most of them the choice is the same. It makes sense: after all, time spent on a beach is Time Out.

What catches my eye, though, as I follow the trail of blue wrappers with their annoying comic-book red lettering, is something a little more unusual: a Marathon wrapper.

As almost anyone of my generation knows, Mars changed the name of the Marathon bar to Snickers in 1990, as part of a move to standardise its products internationally. The change caused outrage at the time and is still surprisingly controversial; there's even a Bring Back Marathon campaign. The organisers begin lightheartedly by saying that Snickers is 'a silly name' which 'sounds a bit too much like "knickers" to be taken seriously'. They go on to claim, perhaps with a little hyperbole, that Marathon is 'part of British history, part of our culture'. Then they get really cross: 'Too many things get chopped and changed by the advertising men. It's crazy and we should all be fighting tooth and nail to stop it.'

The significance of all this to me, as I stand on the beach with the Marathon wrapper in my hand, is that it must be at least twenty years old. This scrap of litter has been around all that time, either in the sea, or buried in the sand. And it's in remarkably good condition: the background colour has mutated from brown to green, but the fabric is undamaged and the lettering is faded but easily legible. That's the defining characteristic of plastic packaging: it's strong and long-lasting. We can obey the instruction to dispose of it thoughtfully, but we can't make it disappear, however often we chuck it away.

The toothbrushes, cigarette lighters, Lego bricks and bottle tops which lie on the virgin shores of Midway have been carried there on a huge vortex current called the North Pacific Gyre. The ocean moves extremely slowly here, and whatever it's carrying gets trapped and can take many years to escape. As a result of these exceptionally sluggish conditions, the Gyre has become home to something known as the Great Pacific Garbage Patch, a gigantic 'stew' of suspended plastic and other human debris. There is contention about the scale of the problem, but estimates put the total volume of the Garbage Patch at a hundred million tons, and it is said by some observers to cover an area twice the size of Texas.

The Garbage Patch was described graphically by the sailor and oceanographer, Charles Moore, who has been tracking it and observing its growth for more than fifteen years. He first observed it from the deck of his yacht on a voyage between Hawaii and Los Angeles. He had steered his boat into the North Pacific Gyre, an area sailors generally avoid because there is so little wind. There, to his horror and amazement, he found himself surrounded by rubbish, despite being thousands of miles from land. He sailed on through it, day after long day, for a whole week. 'Every time I came on deck, there was trash floating by,' he said later. 'How could we have fouled such a huge area? How could this go on for a week?' For Moore, it was a life-changing experience. It drove him to abandon his lucrative oil business and reinvent himself as an environmental activist, campaigning to persuade us to confront the consequences of our collective addiction to plastic.

*

Can we even imagine a world without plastic? What was life like before it so thoroughly colonised our homes, offices, vehicles, streets, shops, gardens and parks? Its success story has been so phenomenal that it's hard to say what we would be without it.

But our grandparents would know. We've only been mass-producing plastic since the 1930s; there was a time within living memory when plastic was a rare sight, and a novelty. The story begins much earlier, in the middle of the nineteenth century, when the home piano was at its most fashionable and demand at its highest, and the very earliest plastics were created to provide a man-made alternative to ivory keys. A new product called Parkesine was demonstrated at the Great London Exposition of 1862; it was enthusiastically received, but it took fifty years for the ready availability of raw materials and the development of production methods to come together and make it a sensation. The last eighty years have been one long uninterrupted boom time for plastic.

The key to its success is there in its name, which comes from the Greek word *plassein*, to mould. Its malleability means that it can be cast, pressed and extruded into an almost limitless range of shapes, and be put to a dazzling variety of uses.

As well as being adaptable, the new wonder-products were also very durable. Back in 1946, when Earl Tupper launched his range of plastic kitchenware and issued the famous Lifetime Guarantee, it must have sounded like a wildly reckless act of largesse. But those sandwich boxes certainly do last; some of the original 1940s pieces are still in use. Nowadays, though, no one's looking for longevity; it's disposability we're after. (Earl Tupper must be turning in his grave: by all accounts, he wasn't one for extravagance.)

Where plastic is concerned, the word 'disposable' has a hollow ring. Since the 1950s, an estimated one billion tons of plastic have been discarded, and most of it will take hundreds or even thousands of years to degrade. At this very moment, every piece of plastic which has ever found its way into the ocean is still out there, somewhere. Ocean currents have swept it in, shuffled it, taken it on long, complicated journeys. They've carried it to the most remote and undeveloped shores in the world, and delivered it onto the clean white sand for the local wildlife to eat.

The oldest piece of plastic found inside an albatross was manufactured in 1944. But some modern plastics are even more durable than older kinds. When photodegradation does take place, the toothbrush or the plastic bottle disintegrates into smaller and smaller pieces, and eventually into minuscule fragments. These fragments are ingested by marine animals, which mistake them for plankton. They enter the food chain. What we chuck in the sea ends up on our dinner plates. If you eat fish, you eat plastic. It's as simple as that.

A major ingredient in the stew is the nurdle. It's a pellet of plastic resin, not much bigger than a grain of wheat; too small, you might think, to matter very much. But there are so many of them. The easiest and cheapest way to get raw plastic from the supplier to the manufacturer of our consumer goods is in pellet form, and billions of these are shipped across the world's oceans every year. But not all nurdles reach their destination. Vast quantities are lost along the way, washed down storm drains in plastics factories or spilt from containers in transit. They end up in the sea, jostling for space with chips of Barbie doll and disposable razor. Each one acts as a tiny sponge, sucking up and absorbing toxic chemicals like DDT. Some have been found to contain dangerous polychlorinated biphenyls (PCBs) in concentrations up to a million times greater than the surrounding seawater.

No one worried about nurdles until a few years ago, but they are now the second most common litter item on beaches, according to the alliance of surfers and wave-riders who are leading the campaign to raise awareness about them. Surfers are right in the front line when it comes to pollution – they're the first to get a mouthful of sewage or oil or plastic, and these experiences have turned the surfing community into an environmental force to be reckoned with. Surfers Against Sewage was founded in 1990 by a group of Cornish surfers who were 'sick of getting sick' with repeated ear, nose, throat and gastric infections. In the early days, their focus was on sewage discharge, but the scope of their campaigning has broadened to include nurdles, which they have christened 'mermaid's tears'.

*

In the twenty years I've lived here, my local beach has cleaned up its act considerably. My copy of the Good Beach Guide, the 1988 edition, featured not one single beach in north-west England, but a huge programme of investment in sewage treatment, along with official designations recognising the special importance of this stretch of coast, brought it up to Blue Flag standard in 2004. A report in the local paper gives the result of a recent Beachwatch survey carried out by the Marine Conservation Society, which concluded that, although nationally litter on beaches is at a record high, beaches on the north-west coast are getting cleaner. Strenuous efforts are paying off. Nevertheless, there's plenty of rubbish on every beach, and this one is certainly no exception.

All things washed up by the tide are altered by it. It blurs the lines between natural and man-made. It's not always easy to tell, as you walk the strandline. Is that a sheep's skull? No, it's a squashed football. Moon jelly? No, polythene kite. Dead fish? Broken-off piece of a bucket.

I set off on an August morning to investigate the strandline and see how much plastic I can find in an hour. A friend who works for the Environment Agency has given me a leaflet about the Garber Survey, a procedure for examining and recording the condition of a beach. The sampler walks three sides of a rectangle, observing and recording on a logsheet all the materials on the List of Determinands:

Dead Marine Life
Animal Faeces
Intact Human Faeces
Grease/Scum Slick
Sewage Debris
Condoms/Tampon Applicators
Sanitary Towels
Noxious Sewage Odours
Terrestrially Derived Litter
Oil and Tar

Each determinand is described in detail and advice offered to help the sampler recognise the items in question, with special instructions in case of particular finds. I learn that 'often only the rubber rings from condoms may be found', and that 'large items such as dolphins and porpoises etc. should be reported to the relevant Pollution Control Officer'.

No dolphins or porpoises today, though there is a dead sheep sprawled on the sand in a rusty puddle. Most of its fleece has been stripped away, revealing skin as tight and glossy as white polythene.

I stuff the leaflet in my bag. This is not going to be a scientific exercise. I'm just going to start at the beach entrance, and spend an hour investigating the high-tide line on the most heavily populated section of the summer beach. And with the albatross and the toothbrush in mind, I'll focus exclusively on plastic.

The strandline is very changeable, depending on weather and tidal conditions. It can be heaped so high you have to wade through it; or it can be nothing more than a sparse trail of twigs and scraps. Today, it's a modest ribbon of debris about a metre wide. A cursory glance reveals just how much man-made material there is, tangled up with the sticks, shells and seaweed. To begin with, I'm surprised by the variety of plastic items I find. After about ten minutes, though, there's very little new; just more of the same, and more, and more. At the end of the hour, I've found and identified a total of 311 plastic items:

Bottle – 5
Bin bag – 1
Pen cap – 3
Pen – 1
Burst balloon – 5
Disposable cigarette lighter – 10
Plastic bag – 9
Binoculars lens cap – 1
Cotton bud stick – 12
Sole of a shoe – 1
Disposable glove – 1
Straw – 9
Petrol can – 1
Drinks carton – 2

Sanitary towel wrapper – 13

Sweet wrapper – 39

Chocolate bar wrapper – 49

Toy dinosaur – 1

Rope – 2

Crisp packet – 4

Bart Simpson stencil – 1

Ice-lolly wrapper – 7

Cup – 3

Tobacco packet – 2

Cellophane wrapper from cigarette packet – 26

Shuttlecock – 1

Length of fishing or kite line – 6

Clothes peg – 2

'Happy Birthday' banner – 1

Document wallet – 1

Miscellaneous fragment – 92

This part of the beach is cleaned regularly during the summer months; I'd have to walk a bit further to find more exotic flotsam and jetsam. Here, it's the height of the holiday season, and this is the site of hundreds of picnics and family outings. On the assumption that much of this rubbish has simply been discarded by beach users, it offers a snapshot of the things people like to do when they go to the seaside. On the evidence available, the unsurprising top favourites are eating confectionery and smoking. Smoking can be a contemplative activity, and aficionados might well find a deeply traditional sort of pleasure in walking at the sea's edge or sitting on the sand gazing moodily out to sea with a cigarette in their hand. It's an alfresco version of the old Parisian café experience:

the sound of the waves and the gulls standing in for the melancholy delights of Edith Piaf. I'm reminded, too, of the old *Players, Please!* advertisements, with the salty sea dog enjoying a hearty lungful of tobacco at the end of a hard day splicing the mainbrace.

I found the document wallet and the pen some distance apart, but I'm sure they must have arrived together. I imagine them carried down here in a briefcase, businesslike and anomalous. Its owner has caved in to pressure from partner and children, but clings still to an insistence on doing a bit of work, perhaps wishing to make a point ('It's all right for you; I'm far too busy to take time off!'). I like to picture this wage slave attempting to deal with important paperwork in the essentially anarchic surroundings of a windy beach, with blown sand and an incoming tide and children and dogs and kites and frisbees. Chasing crucial documents as they skitter away on the breeze, and finally giving up, collapsing with mingled terror and relief onto the sand. Perhaps unwrapping a chocolate bar, perhaps sparking up a Marlboro. Lying back at last and watching the clouds. Letting it all go.

But then there are the sanitary towel wrappers, fluttering in the breeze in their bright array of purples, blues and greens, like little flags planted in the sand. What we flush down the toilet ends up in the sea; but do people really flush plastic wrappers? The screens and grilles and raking systems employed by the sewerage industry trap most large items before they can escape from the pipes into the sea, but these wrappers, and the plastic sticks from cotton buds, are notoriously difficult to catch.

⋆

The albatross is an impressive creature in real life; it has iconic cultural significance too. Thanks to Coleridge and his *Rime of the*

Ancient Mariner, the albatross has become a metaphorical burden, usually worn around the neck. Ironically, though, such significance has been a heavy burden for the albatross to carry. These birds have been hunted, not only for food but also out of fear and superstition. The albatross has been perceived variously as a guide, a harbinger of change in the weather, a good omen, a bad omen, a heavenly messenger and an evil spirit.

In the poem, there is confusion amongst the ship's crew in the aftermath of the killing. They can't decide what to think. At first they are angry with the Mariner, because they believe the albatross was responsible for the favourable south wind which has brought them out of the Antarctic. Then, when the persistent fog lifts, they change their minds and congratulate him for getting rid of this curse which brought them bad weather. When the wind drops and the ship is becalmed, they turn on him again, hanging the dead bird around his neck as a sign and a punishment.

The world view of Coleridge's sailors is one in which man is locked in a power struggle with nature; the old gods can be angry and vengeful, and must be appeased or vanquished. To our contemporary eyes, the albatross is caught in the middle, a victim of ignorance and superstition, held to account for every change in the weather. In reality, it too is subject to the unpredictable fortunes of wind and waves; it's built for soaring, not flapping, and it can be 'becalmed', just like a sailing ship, and forced to rest on the ocean's surface until the wind picks up again.

When the poem was first published in 1798, it divided readers and critics with its mixture of archaism and prophecy. On one possible reading, it carries a stark warning: interfere with the natural order – break the web – and appalling consequences will follow. On another, it plunges the reader into a dark and decidedly twenty-first-century vision of the randomness of a godless universe, its meaningless accidents of death and disaster. Either way, contemporary readers can be surprised by its very modern morality, and the prefiguring within it of what we would now describe as environmentalism.

The albatross was a symbolic creature long before Coleridge wrote about it. Because these birds would follow ships throughout long voyages, and were thought to embody the souls of dead sailors, they were regarded by generations of seafarers with reverence and even a degree of fear. But the tedium of life at sea was sometimes relieved with a kind of sport which consisted of baiting and trapping them, and holding them captive for a while on board ship before releasing them. In his poem 'L'albatros', Baudelaire describes one of these 'rois de l'azur', captured and brought down onto the deck. What was magnificent in the air looks very different lying helpless at the feet of the crewmen:

Torn from his native space, this captive king
Flounders upon the deck in stricken pride,
And pitiably lets his great white wing
Drag like a heavy paddle at his side.

This rider of winds, how awkward he is, and weak!
How droll he seems, who lately was all grace!
A sailor pokes a pipestem into his beak;
Another, hobbling, mocks his trameled pace.

Baudelaire goes on to draw a comparison with the poet, out of his element in the everyday world where he is ill-equipped and misunderstood: one moment he's 'monarch of the clouds', the next 'exiled on earth', hampered by his useless wings, an object of mockery. It's an uncomfortable double stereotype: poet as unworldly creature in possession of rare and special gift; poet as figure of fun. But it's the the downed albatross that occupies the centre of the poem: a more starkly powerful image than ever now, in this age of death by plastic.

★

In her book *Findings*, Kathleen Jamie describes an expedition to Ceann Iar, a remote and uninhabited Hebridean island. In this wild place, she and her companions are astonished to find – along with seals and rare seabirds, a dead whale and pieces of a crashed aeroplane – a huge quantity of plastic, driven onshore by the wind and waves and trapped between the sand dunes. After a while, though, nothing shocks them any more; the bizarre variety of objects starts to seem almost normal:

> Here in the rain, with the rotting whale and wheeling birds, the plastic floats and turquoise rope, the sealskins, driftwood and rabbit skulls, a crashed plane didn't seem untoward. If a whale, why not an aeroplane? If a lamb, why not a training shoe? Here was a baby's yellow bathtime duck, and here the severed head of a doll. The doll still had tufts of hair, and if you tilted her she blinked her eyes in surprise.

We are accustomed to thinking of the oceans as wild, unspoilt parts of our planet. Their depths are still largely unexplored, and some of the rarest and most enigmatic creatures in the world roam freely there. On aesthetic grounds alone, images of plastic bags and bottles washed up on remote island shores are deeply disturbing. The seaside enjoys special status in the collective imagination; we have thought of it as a healthy place. Many of us 'get back to nature' by swimming in the sea when we're on holiday, and we love to watch the tide come in and wash the beach clean. Back in the 1950s world of *The Arrow Seaside Companion*, Hugh Stoker's jolly chapter on 'Camping by the Sea' extols the straight-forwardly wholesome pleasures of gathering shellfish from rock-pools to cook in a pan of seawater, with never a note of anxiety about sewage waste, radioactive contamination or sinister sex changes in limpets. As a species we have always had a powerful emotional relationship with the sea, and to learn how comprehen-sively we are trashing it is to experience an agonisingly painful loss of innocence.

The trouble is that the language doesn't reflect the reality. We talk about 'getting rid' of what we no longer want; the original meaning of the word 'rid' was to clear, but because plastic is so durable we can never really be clear of it. It lives on in the system, turning up on this or that beach according to a logic beyond our own. We 'throw things away', but we can never throw them far enough; the sea always brings them back.

It brings them back, and takes them away, and brings them back again. My most sobering moment on this beach was not one spent picking through trash in the strandline as I did today. It was a glorious day in March, after the high tides of the spring equinox. The sea had come in much further than usual, right into the dunes, and washed the beach clean and shining. The mass of accumulated debris I'd seen there the week before was gone. The sea had swallowed it again. I understood then that for the bottle and the laundry basket, the clothes peg and the doll, the petrol can and the chocolate wrapper, this is a journey with no end.

Swarm

When I was a child, it was said to be lucky to have a ladybird land on your sleeve. My dad certainly liked it when they visited the garden, because they ate the greenfly that plagued his raspberry canes. There was an air of mild enchantment carried by the single ladybird – the one that chose you and rested on you, perhaps just for a few seconds, perhaps for a minute or two, before unfolding itself and flying away home. In some places, they are known by odd and affectionate nicknames, like *dowdy-cows* and *lamb-ladies*.

One muggy August day, there is a phenomenal swarm of seven-spot ladybirds on this coast. Millions of them divert onshore, interrupting their journey to and from who-knows-where, in response to some rubric laid down by recent days of rain and high temperatures. They're bizarrely out of place here on the beach. They fill the air, and seagulls make slow predatory circles overhead. They blunder drunkenly on the ground, turning the sand red. They drip from fenceposts and pool beneath.

We walk on the beach with our faces covered in scarves, though it's midsummer. Still they land on our eyelids and in our hair, and tumble ticklishly down inside our clothes. As we walk, they crunch underfoot; they're everywhere, it's impossible not to tread on them. Families abandon their picnics and flee. People lock themselves inside their cars, with the windows wound up in spite of the heat. The windscreens are alive with ladybirds; I watch one driver try to clear them, but the wipers can't work fast enough.

These childhood gifts, these small enchantments, are suddenly so extravagantly numerous that their currency has collapsed. They've become worthless, and worse than worthless. Now they look instead like tiny curses.

They've probably been drawn here by chemical attraction – they're acutely sensitive to pheromones given off by aphids, and also to their own home-grown version. They lay pheromone trails for one another, leading to prime hibernation sites in winter. The harlequin ladybird, larger and more aggressive, a relative newcomer to these shores, is particularly good at this; if you get one hibernating in your house, you get thousands.

Swarm behaviour is purposeful, but the members of the swarm can be surprisingly unfocused. Together, they seem to possess a collective will, but when I stop and watch a few individuals their movements look vague and pointless. They ramble indecisively in the air, or mill about on a driftwood log as if lost and amnesiac.

What is it about swarms that raises the hair on the back of our necks? Ladybirds are not harmful: they don't sting like wasps, and they don't devastate crops like locusts. In fact, they have a voracious appetite for the aphid, which is regarded as a garden pest. But it seems we can have too much of a good thing. A sudden onslaught on this scale feels threatening; it feels like a plague. Things are out of kilter. What does it mean? What will come next? Is it an omen of some kind; a judgement on us? In biblical times, people read into events like this divine warning or vengeance; now it's our transgressions against the environment, our hubristic disregard for the natural order of things, which preoccupy and terrify us.

'Ladybird Books,' laughs my friend. 'Remember their slogan?

Everybody loves a . . . ' He stops abruptly and spits one out. The beach is almost deserted by now, and this is starting to feel apocalyptic.

A dog walker coming the other way calls out: 'Aren't they horrible little bastards?' And he breaks into a run.

Old Seafarer

In his prose poem 'The Dead Seal near McClure's Beach', Robert Bly writes of an intimate encounter with a seal which at first appears dead but is actually in the long, slow process of dying. It's lying still, breathing almost imperceptibly, until he reaches out to touch it:

> Suddenly he rears up, turns over, gives three cries, Awaaark! Awaaark! Awaaark! – like the cries from Christmas toys. He lunges toward me. I am terrified and leap back, although I know there can be no teeth in that jaw.

However I might wish it, the seal I'm looking at this afternoon is not going to spring to life like that. It's dead beyond a shadow of a doubt, already bloated and giving off a penetrating stench.

It's by no means rare for a dead seal to be washed up on the beach. Often there is no obvious cause of death, though there has been a spate of gruesome incidents recently. Dozens of seals have washed up on Scottish and English beaches with fatal 'corkscrew' injuries: spiral-shaped lacerations from head to tail. An investigation has found that the wounds are caused not by Greenland sharks, as originally suggested, but by the ducted propeller systems or azimuth thrusters on ships operating in shallow coastal waters. Ducted propellers are used to maintain the position of a ship, by altering the speed and direction of their thrust. The vessel is idling, almost stationary; the seal approaches. Then suddenly the thruster starts up, or reverses its rotating propellers, and the seal is sucked into the system. There's no way it can escape; the force of water driving it between the angled blades is so great and so irresistible, it's capable of tearing skin and blubber right off the underlying muscle and skeleton.

I'm glad never to have found one of these appallingly mutilated creatures. This one has no obvious injuries, except for pale scarring like patches of blistered paint on its side. It looks like a venerable old specimen, hauled out by age rather than anything more sinister.

*

The Latin name of the Atlantic grey seal, *Halichoerus grypus*, can be translated as 'hooked-nosed sea pig', and the 'Roman' shape of the nose is one of the features which distinguishes it from the common seal.

Still, the name hardly seems to do justice to this handsome animal with its brown and silver-grey dappled coat. In summer grey seals can be seen here, swimming and 'bottling' – hanging vertically in the water with their noses pointing to the sky. It's easy from a distance to mistake the bobbing head of a grey seal in the water for the sea-sleek head of a human swimmer.

They bring a touch of wildness to a domesticated day at the seaside; but they are not shy. They will swim up quite close to the shore, and if you stand at the water's edge and look at them they will return your gaze, steady, curious and unafraid. It's easy to imagine one of these watchful creatures slipping out of its skin and joining the human world for a time, like the legendary selkie.

Seals can live to be thirty years old, so it's not inconceivable that this one was born when I was a teenager. That was a time when seal culling – and protests against it – were much in the news. I remember graphic television pictures of pups being clubbed to death. They huddled defencelessly, gazing out of the television screen through huge eyes. The dead seal at my feet might well be a contemporary of those pups.

The grey seal is now protected by law, but the same arguments rage as bitterly as ever. The fishing industry claims that seals take hundreds of thousands of tons of valuable fish a year. They are not fussy eaters – they're as happy with sand eels as they are with cod – but in the context of diminishing stocks and fishing quotas they are regarded by some fishermen as a threat to their livelihood.

Environmentalists are divided. The evidence is open to interpretation, and the pro-cull lobby has been accused of using the seal as a scapegoat for bad management of fisheries. But the seal population has risen steeply since the ban, and it's possible to make a case for controlling numbers in order to balance the needs of other species.

There's no doubt, though, that the general public finds culling abhorrent. People are very fond of seals. They are one of the few really large wild animals most of us ever get to see. Their wildness matters very much to us. And of course there's nothing more photogenic than a seal pup. It's the very epitome of vulnerability, deposited on open ground with no burrow or camouflage to protect it. As the naturalist John Lister-Kaye puts it, 'our Western culture loves all things young, furry and helpless'.

*

The carcasses of seals, sheep and other animals which end up on our beaches are a bit of a headache for the local authorities. These animals are usually taken for burial in landfill sites, but since a grey seal can weigh as much as 250 kilograms, it's a logistically challenging task. I've seen workers attaching ropes to a dead seal and tying it to the back of a Land Rover in order to drag it up the beach and away.

Dealing with it quickly is desirable, of course – once scavenging birds and animals have begun to take advantage, the integrity of the corpse is broken and the job becomes more difficult. Visitors are liable to complain about seeing a dead animal on the beach, and they're even more upset if it's in a scattered and half-eaten state. Putrefaction sets in, and the smell of a dead seal is over-powering after a few days. There's also the danger of a build-up of gases, bloating the corpse and presenting the risk that it will burst when you attempt to move it.

If the body is on your land, you have to deal with it yourself. Jeremy Clarkson, writing recently in the *Sunday Times*, admitted wryly that this sort of thing was not quite what he'd expected when he moved to an idyllic cottage by the sea. Faced with a dead seal uncomfortably close to home, he tried with staggering ineptness to refloat it, burn it and bury it, all to no avail.

Annual migrations to and from the breeding grounds mean that in its lifetime a seal will travel hundreds of thousands of miles. Researchers studying migration are able to identify individuals, because each one has a unique fur pattern, like a human fingerprint, which it keeps all its life. I can't read the language of fur patterns, but I can guess that this individual probably came here from the colony at Hilbre Island in the Dee estuary. Hilbre is a nature reserve frequented by birdwatchers, who visit for the terns, oystercatchers, curlews, purple sandpipers and redshanks. The grey seals are another attraction: up to five hundred of them spend spring and summer swimming and hauling out on and around the island, before returning to their rookeries in the autumn.

This swollen old seafarer in his shabby overcoat has come to the end of his travels. But standing over the carcass, holding my nose against the smell, I suddenly recall a luminous childhood experience. I was about eight years old, swimming off West Dale beach in Pembrokeshire. I surfaced from a dive, to find that all my playmates were scrambling and running out of the water.

I turned around and saw it, just a few feet away. It was looking at me, and for a long, breathless moment, I trod water and looked back. The polished head, almost motionless on the surface. The obsidian eyes.

Autumn

Drifters

On a remote threshold at the top of the beach, I find a door. A few purple heads of sea holly grow around it like the remnants of a cottage garden, each head wickedly sharp and wearing a wide starched ruff of leaf skeletons. The door is installed flat on the sand. Its blue paint has been scrubbed by the tide and bleached by the sun. There's no handle, so I kneel and press my eye to the keyhole instead. It's dark as a locked library in there.

A door is a very particular and functional thing, rarely encountered anywhere except hinged to a frame in a building. You might see doors stacked in a timber yard or DIY shop, like the warehouse on the outskirts of Liverpool intriguingly called World of Doors and Fires. They are seen too in skips and rubbish dumps. But to find one here is strange, unexpected, a reminder of the indiscriminate way the sea takes things in and then throws them back. This door managed to escape its role as a divider of domestic spaces and has spent weeks or months adventuring in the open sea before being flung out by the tide to lie warped and faded on the sand.

Here on the beach, the usual significances are lost, the ordinary object is stripped of context and the familiar made strange. Earlier this afternoon, I stood and watched as the tide came in – one of those exceptionally gentle, almost silent tides we sometimes get here – and on its surface a bright acid-yellow workman's glove slithering up the beach like a luminous severed hand. A little further on, there was a plain glass bottle, scoured clean of its dubious contents, label gone. It looked glossy, brand new, innocent. The detritus of our lives is washed, softened and given back to us cleansed of its dirt and shame. That's the work of the sea. It comes in faithfully, on schedule, like an old-time religion, and washes away our sins.

This door may have been discarded from the shore, or from a cargo ship – who knows? For as long as humans have been making objects out of wood, we have also been losing them, and throwing them away; and what is thrown away has a habit of finding its way into the sea, and ultimately onto our beaches. It's not just boomerangs that find their way back.

The sea carries a huge variety of wooden debris, the greatest proportion of it unworked and naturally occurring. Just a few minutes before I found the blue door, I saw a giant black spider which turned out to be an enormous branch, dense and brittle with salt, with wisps of seaweed caught on the twigs like bedraggled leaves. Another time there was a dark mass in the distance which I thought might be a stranded seal or a dead sheep but was in fact a whole tree trunk, at least twelve feet long, with part of its roots still attached, like hair growing on the head of a corpse in a Gabriel García Márquez story. It was blackened by salt water, cured to the hardness of bone. I've no idea how far it had travelled, or how long its journey had taken, but it was massively heavy. Just imagine the force required to carry this tree trunk, and to cast it up on the sand. But this is nothing compared with sweeping a house off a cliff or dashing a ship to pieces.

*

When I lived in Margate as a student in the early eighties, driftwood saw us through a cold winter. The summer beach, crowded with day-trippers in rented deckchairs, was not for us, though we did visit Dreamland, the town's legendary seafront funfair, which boasted, amongst other pleasures, the best doughnuts I've ever tasted before or since. Like Proust's madeleine, those doughnuts bring it all instantly to mind: that place, that vivid year of my life.

But it was in the winter that we began to explore the beach, and to scavenge there. Our student house, little better than a squat, had two or three broken windows, and a tramp climbed in at night and slept in the bath. There was a prolonged cold spell before Christmas, and thanks to the broken window I woke up once or twice with a sheen of frost on my blankets. For heating and hot water we had only a knackered old Rayburn stove, and it had to be fed somehow.

There were streets and streets of these Victorian terraces in Margate, largely unmodernised and each with its original cellar and coal hole from the street above. To begin with, we started a kitty and paid for coal to be delivered, but our student grants were quickly spent on other things and we resorted to hiding when the coalman called for his money. I can't remember whose idea it was, but one bitter December evening, as we sat shivering and clapping our arms like Richard E. Grant and Paul McGann in *Withnail and I*, someone remarked that there was in the town a source of fuel, free of charge and there for the taking.

Next morning, in the biting cold, we set off en masse on an expedition to the beach with ropes, attached them to the largest pieces of driftwood we could find, and dragged them back through the streets to our house. Perhaps the neighbours didn't like it much, though ours was not what you'd call a *good area*, and in any case we were young and oblivious. All that winter, we curated, amongst the half-bricks and broken bottles in our overgrown garden, a collection of these weird wave-sculptures. There would always be a couple more drying out very slowly on the kitchen floor, blocking the way to the fridge, hacking shins and tearing sleeves as we passed, filling the house with the smell of salt and decay.

*

Driftwood forms the very stuff of Creation, according to Norse mythology. Ask and Embla, the first humans, were originally pieces of driftwood found on the beach by Odin and his two brothers. The three young gods breathed life into them and endowed them with vision, speech, intelligence and other human gifts. Like Adam and Eve, Ask and Embla are the ancestors of all human beings, yet their beginnings are as simple and commonplace as the tree trunk I'm looking at today.

Gods are not the only ones. Children too understand the creative potential of driftwood. They make seesaws and rockets and dens from it. In a letter to the *Guardian* following its obituary of Beryl Bainbridge in 2010, Michael Vaughan-Rees recalls interviewing her, gathering her reminiscences about this very beach and the time she spent here on childhood outings from Liverpool in the 1940s. She devised a popgun made from a piece of driftwood, and another piece on a length of string became a pet dog, which she named Blaze. 'I'd go out barefoot, then cross the road to where I kept my brother's old cast-off trousers hidden in some bushes, take my gymslip off and change into the trousers. Then I'd be away, the dog in one hand, my rifle in the other, and just go "Bang!Bang!" at anything in sight; hour after hour, it was lovely.'

It may look ordinary, but its exceptional status in the Creation myth is a reflection of its unique value. Wood means fire, and fire means warmth, food, survival. Before the availability of other sources of fuel, it was a vital and indispensable commodity, and this is surely one of the reasons why trees have been regarded by so many societies as sacred. Not only are they tall and rather humanoid in shape, not only do they provide fruit and shelter and other resources; they are precious too after their death because of their capacity to make fire and to provide building material. No wonder they are associated with magic and inhabited by spirits. No wonder tree worship persists, with people still hanging charms and tokens on the branches, and hammering coins into the bark. I've seen one such 'wish tree' on the riverside path at Dovedale in Derbyshire, with a substantial hoard of pennies and shillings embedded deep in its trunk. Like tossing coins into water, this practice has ancient religious origins; we might try to laugh it off as a bit of fun, but these coins are votive offerings to the gods.

For practical reasons of survival, people who live near coasts have always regarded driftwood as a precious resource, especially in places where there are few trees nearby. In the open-air museum at Skógar in Iceland – a country with little in the way of forest – you can visit an old house, once typical, made entirely of driftwood. It was collected and used as a building material here in England too. But perhaps the most enterprising users of it are the Aleut people of south-west Alaska. Just about every artefact you can think of they have made of driftwood, including kayaks, bows and arrows, snowshoes, dog sleds, house frames, fish traps, masks, dolls, model boats and kitchen utensils. They even made bentwood hats, and wooden snow goggles. Times have changed, and some of these items are no longer needed, or are mass-produced and imported these days, but the fundamental significance of driftwood in Aleut life remains. Many of these people have little or no access to growing timber, so the wood which arrives on their shores, brought there by the combined forces of wind, tide and current, is an exceptionally important harvest. As well as making things from it, they use it for the traditional fire-bath, for the sauna, for smoking fish and heating their homes.

★

My own youthful experiences taught me that driftwood, no matter how dry it looks, has a high moisture content and takes a long time to dry out before it will burn well. It can also be very difficult to cut; it may appear clean, but there's sand embedded in every crack and crevice, and it blunts your saw. When you burn salty wood, you get fantastically colourful flames and lots of spitting and crackling, all of which makes for an exciting fire. True, there's the alarming array of chemicals which may be present in timber drifting on the ocean or piled on beaches: all sorts of oils, tars, creosote and other preservatives, which are probably far from wholesome when you toss them onto the fire. But walking on the beach now, I'm reminded of the smell of wood drying in that awful kitchen in Margate: how comforting it was, how it carried the promise of warmth and safety.

Our foraging on the beach for fuel was necessary, harmless, even enjoyable. The work got us warm for a few hours, and the wood did the same for a few days afterwards. There's something immediate and satisfying about gathering wood; a direct link is made between that simple work and the simple pleasures of eating and staying warm. The elemental business of fire making, which is in danger of disappearing from everyday life, still has a hold on us emotionally; everyone likes a 'real fire', and I suppose the popularity of the barbecue is a relic of our primitive relationship with fire – how else can we explain the inconvenience, the mess, the chicken and sausages simultaneously charred and undercooked?

Here on the beach and in the sand dunes, on summer evenings, people make fires out of whatever wood they can find. It's against the local bye-laws, but they do it anyway. Occasionally they bring food to cook, and sometimes they sit around in a circle and drink. Blackened places are what's left of these festive gatherings next day: a small heap of rain-soaked cinders, often strewn with cans or bottles. Occasionally people have given up on the search for driftwood and used whatever else they can lay hands on: there'll be a raw hole in the sand where a signpost has been rooted out like a tooth. Further along the coast, where it's possible to visit at low tide a fragment of petrified forest, it's said that some of these prehistoric tree stumps have been dug out and carried away for firewood.

However, with the luxury of central heating and the ready availability of building materials, more and more driftwood is left to collect on our beaches. In some places it's so abundant that it becomes a nuisance. But it can also become a source for artists, in combination with other recovered materials like limpet shells, stones and dried seaweed. The sea itself makes its own strange artefacts, combining rope and wood and miscellaneous plant material into the eclectic tangles known as 'mare's nests', a name which describes their muddle and untidiness as well as their illusory and paradoxical nature. The beach is often dotted with them, ticking as they dry out in the sun, frequented by insects and picked over by opportunistic gulls.

And that's the thing: nothing is wasted. It may look like debris, but driftwood is a source of shelter and sustenance to a whole host of small shoreline creatures such as crabs and sandhoppers. Before it gets here, some of it has provided life support for the shipworm too.

The shipworm – not in fact a worm at all, but a type of mollusc – lives on nothing but submerged timber. Its ability to survive by burrowing through wood was a mystery until the nineteenth century, when a French geologist described a unique organ now known in his honour as the Gland of Deshayes. This gland produces a bacterium which enables the shipworm to digest cellulose and harvest nitrogen. This is the secret of its success.

A hundred years later, the Gland of Deshayes was unexpectedly in the news again, when the bacterium was hailed as a new miracle ingredient in biological washing powder. It works by generating an enzyme that breaks down cellulose and digests other proteins such as those in bloodstains or spilled milk. What's more, it works at low water temperatures and doesn't cause environmental damage like 'artificial' enzymes. Powerful and eco-friendly: the humble shipworm has it all.

This creature has not always been so popular. It's notorious for the damage it does to wooden ships, piers and docks. The larvae invade the wood, and once grown they start to burrow, using the shell on their heads to cut through, turning this way and that to make a perfectly circular tunnel. A white, chalky material they secrete at the same time lines the tunnel on the way through. Driftwood full of telltale holes is a common sight on the beach, but a shipworm infestation often goes unnoticed until the timber is riddled with tunnels, and eventually the bridge or dock becomes so badly weakened that it collapses. This is still a major problem for structures supported on wooden pilings, and for marine archaeologists studying underwater shipwrecks. But the development of modern materials like fibreglass means that modern-day mariners are less likely to be marooned on Jamaica, as Christopher Columbus was in 1503, because all his ships had been eaten by shipworm.

★

Everything has to eat something, and everything is used, some-times in unexpected ways. Wood drifting on the surface of the sea has played a defining role in the way species are distributed across the planet. Evolutionary biologists now hypothesise that Madagascar's extraordinarily diverse range of mammals arrived there from continental Africa by sea, floating on natural driftwood rafts, rather than by land bridges as previously thought. The animals hitched a lift across the Mozambique Channel, carried by storms and currents, in a mass migration that went on for millions of years. This kind of 'island hopping' has been theorised about for a long time, but the advent of DNA analysis allows zoo-geographists to see in detail how different species are related and to test these theories against the evidence. It now seems likely that the giant tortoise, the most famous symbol of the Galapagos Islands, first arrived there clinging to a piece of driftwood from a river mouth along the Pacific coast.

We too have found numerous and varied ways of using not only wood but also the creatures that live off it. To people in the Philippines, shipworm is an edible delicacy, eaten raw with vinegar or lime juice, chopped chilli peppers and onions. But even those of us who have never tasted it have reason to be appreciative. The shipworm's voracious appetite and expert tunnelling skills are very important in the wider ecosystem; without them, the oceans would be full of dead wood. Their recycling efforts are assisted by the gribble, a smaller creature with a passing resemblance to the woodlouse. It exists quite happily with the shipworm, because it's only interested in the surface layers of timber, where it bores more delicate burrows punctured with tiny ventilation holes. I've never heard of anyone eating gribbles, but they too have succeeded in catching the scientific imagination, this time with the suggestion that their digestive enzyme, copied in the lab, might be used in the process of producing alternative fuels from biomass such as wood or straw.

They are viewed with ambivalence, these minuscule animals. Their destructive potential seems out of all proportion with their size, and through the centuries we have waged war on them with creosote and copper and whatever else we can lay our hands on. Yet we have started to find value in exactly those things which make them so destructive. We never stop looking for ways of using our fellow creatures; we're always sizing them up for food or fuel or the next bit of technology.

*

The tide is coming in and leaving me marooned on a dry spit of land. I look back, and the driftwood 'spider' is already submerged. The door will be next. The sea surges into all the hollows, and two bodies of water inch towards each other to close the dry gap. The next haul of twigs and tree trunks, fenceposts and chair legs is on its way in.

From its role in shaping global biodiversity to its place on the beach barbecue, driftwood is commonplace but powerful stuff. For me – a beach-walker and a fire-maker – it still has some of the old magic. In my imagination, it smokes bleakly against an ominous sky on the shore at Viareggio, in Fournier's painting *The Funeral of Shelley*, with the shocked circle of friends standing by as the body of the drowned poet is burned.

Or it crackles with significance in the grate, like 'The Fire of Drift-Wood' in Longfellow's poem of this name, from *The Seaside and the Fireside* (1849). This fire, 'built of the wreck of stranded ships', leaps into life and then subsides as the sea wind blows

through the old Massachusetts farmhouse and calls forth fiery
secrets of another kind between the people sheltering there:

> The windows, rattling in their frames,
>> The ocean, roaring up the beach,
> The gusty blast, the bickering flames,
>> All mingled vaguely in our speech;

> Until they made themselves a part
>> Of fancies floating through the brain,
> The long-lost ventures of the heart,
>> That send no answers back again.

Sea Potato

'Rough hollow pearls in the seaweed night'. This is how the poet David Bateman describes sea potatoes, which litter this beach at all hours, whatever the weather. I've rarely visited without seeing a sea potato.

It's a pale oval, the size of a pigeon's egg, with a deep groove at the broad end, and marked with notched, curved lines like hand-stitching on a leather shoe. Sea potatoes are usually found well up near the top of the beach, because they're so light and easily caught by the wind. Often they shatter as the wind bowls them along the hard sand – they are very fragile, thin and crisp as wafer. Many times I've picked up an especially pretty specimen and taken it away with me, only to turn out a pocketful of crushed fragments when I get home.

These delicate mementoes are the tests of the heart urchin. The word 'test' comes from the Latin word for 'tile' or 'shell', and it's the urchin's endoskeleton, made of tiny chalky plates fused together. This is what's left of the creature after death, and they are cast up in their thousands here most days. Occasionally there are dead urchins, too, still covered with soft felt, rather than the spines you might expect.

Sea urchins are so called because they are said to resemble hedgehogs – or 'urchins', to use the old country word – by virtue of their shape and their spines. The heart urchin is an 'irregular' species, and although it shares the characteristic fivefold symmetry, it has shorter and finer spines than its relatives. It differs in another way too. Regular urchins chew their food using a distinctive bony jaw known as Aristotle's lantern. The name is derived from the *History of Animals*, written two and a half thousand years ago, in which Aristotle gives a meticulously detailed description of the physiology of the jaw, and likens it to 'a horn lantern with the panes of horn left out'.

The heart urchin neither has nor needs a lantern; it feeds in a simpler way. It spends its life in burrows a few centimetres below the surface of the sand, making a narrow hole in the roof to let in clean, oxygenated water for respiration. As they dig, its tube feet gather up organic material and pass it to the mouth for feeding. Fine hairs around the mouth trap the tiny food particles and pull them in.

The heart urchin may miss out on the distinction of having its jaw described by an Ancient Greek philosopher. But when it comes to longevity, it wins out over the rest. No matter where they are in the world, regular urchins have too many predators to count: sea otters, sea stars, crabs, eels, lobsters, wrasses and triggerfish, amongst others. By contrast, the heart urchin is relatively unpopular as a source of food, and can survive for up to twelve years – a grand old age for a small marine animal.

But it's in life after death that it really breaks records. Because it lives in the soft sediment, it's more likely to make a lasting impression, and fossil heart urchins can be found on rocky shores in southern England, dating back to the Cretaceous period, eighty million years ago.

In some places these fossils, immaculately preserved in flint, are so numerous and so striking in appearance that through the centuries local people have collected them, and given them special names and associations. At times they have even taken on a magical significance, and it's not hard to see why – they are exquisitely beautiful with their symmetrical rays and patterns, and they must have been deeply puzzling to people digging for flint to make tools. They were used as amulets and jewellery, and in one Bronze Age burial site a hundred of them were found, ceremonially arranged in a circle around the skeletons of a woman and child. In rural areas, they were until quite recently known as shepherd's crowns or fairy loaves, and it was common to collect them and keep them on the hearth or cottage windowsill. They were domestic charms, bringing good luck of various kinds and helping to predict the weather.

In their fossil life, transformed with flint, they are strong and durable things, full of potential, talismanic. But cast on the beach, they are all emptiness. The incoming tide lifts them effortlessly and sends them skittering up the beach, light and free as blown froth.

Come in, Number 189

An invasion of ducks was predicted for the south-west coast of England a couple of years ago, but we're not expecting any here. Besides, the moment I spot this duck I know for sure it's not one of the famous cargo of Friendly Floatees bath toys which spilled from a container ship into the Pacific Ocean in 1992. The ducks and beavers from that consignment are now bleached white after so many years in the water (though curiously the frogs and turtles have kept their original colours).

This duck, tucked away shyly in the shade of a large driftwood log, is still bright yellow, and it has a number painted underneath: 189. This individual must have been taking part in a duck race down a river, back in those golden days of summer which have somehow given us the slip. Perhaps it was the centrepiece event of a village fete or primary school fundraiser. But whoever placed their £2 bet on number 189 went away feeling cheated. Number 189 never breasted the finishing tape. It had other ideas. It slipped free of the jostling flock as they bumped over stones and snagged on fallen branches. It broke away, following its beak, finding a different channel, a swifter current, which swept it down some small, forgotten tributary overhung with willow and alder, and the voices of the crowd faded and were lost. For 189 had escaped.

★

All rivers eventually find their way to the sea, and the journey changes again. Powerful currents and tidal systems take hold. The twenty-eight thousand or so bath toys lost overboard in the Pacific have given oceanographers an unusual opportunity to track these systems. This work has been led by Curtis Ebbesmeyer, who co-ordinated a worldwide network of volunteers to comb their local beaches and report finds. The first toys – hundreds of them – washed up on the coast of Alaska. Others took a more southerly route to Australia and Indonesia, while an unknown number floated towards South America. An especially adventurous contingent went north, navigated the Bering Strait, entered the Arctic Ocean and became trapped in pack ice; it could take years for them to edge their way to the Atlantic, where the ice will thaw and release them.

So far none of the ducks has turned up on the British coast, though there have been several false claims. People really want to find one of these toys; they have become highly collectable items, and their story has taken on a cultish significance across the world.

In his book *Flotsametrics and the Floating World* (2009), Ebbesmeyer explains the science behind the journeys taken by ocean debris: bath toys, driftwood, messages in bottles, corpses and derelict ships. Before the ducks came along and grabbed the media spotlight, he was already studying other cargo losses, like the thousands of Nike training shoes which he refers to as 'the Great Sneaker Spill'. The shoes were not paired for transportation, so they washed up singly; but beachcombers in Oregon collected them, scrubbed them clean of algae and barnacles, and held swaps to match up pairs. Unlike the bath toys, they were expensive items, so it was well worth the effort.

These stories are making me revisit and reinterpret some of my own beach finds. I've often noticed a kind of 'rule of recurrence':

I find something unusual – something I've never seen here before – and almost immediately I find another the same, and then another. And certain kinds of object come and go: they're numerous when I visit one week, and have vanished by the next. It looks like something more than coincidence.

A peculiar example of this is tins of Nivea face cream. I saw one lying on a bed of twigs in the strandline, and photographed it. I was curious about it, because it was so obviously a very old tin, of the sort my grandmother might have had on her dressing table: you could tell by the obsolete design (white with a blue stripe), the style of lettering, and the old-fashioned spelling of the word 'crème', with all its sophisticated continental associations. *For the care of the skin* it said, in case there should be any doubt. I tried to open it, but it was welded shut with rust.

I left it and walked on. Five minutes later, I found another old tin, absolutely identical. And then another. I counted seven, within a few hundred metres.

How did they all come to be here, on the same day? Could they have been washed off a cargo ship? And if so, when? Flotsam has been known to get trapped inside ocean gyres for fifty years. Is it possible that these tins have been making slow circles out there since the days when I was visiting my gran and exploring her dressing table?

*

The ducks may be children's toys with cute faces, but they're still plastic, which means they're not really going anywhere. There are still thousands of them circumnavigating the globe and washing up on the world's emptiest shores.

The sheer scale of the original spill hints at our massive

145

over-production of plastic objects. The twenty-eight thousand were the contents of just one container, and modern ships are capable of carrying up to fifteen thousand containers. Can there really be enough demand from clamouring toddlers for such colossal numbers of plastic bath toys?

The truth is, they're sold for almost nothing, and designed to be more or less disposable. But they're tough, too. They are designed, the manufacturers have claimed, to survive fifty-two dishwasher cycles; that explains why they're still being washed up in reasonable condition, after nearly twenty years in the sea.

A pack of three new-style Friendly Floatees is now advertised for the modest price of £3.97. They don't resemble the ones lost overboard; in fact, it's a little difficult to tell what the three stylised figures are: the yellow one looks more like a seal than a duck, and the green one is vaguely ursine in shape. I suppose the babies for whom they're designed probably don't mind too much whether the ducks are anatomically correct. At the time of writing, the Amazon website lists them as currently unavailable, and says 'We don't know when or if this item will be back in stock.'

There's no shortage of alternatives. For the right price, I might even part with Number 189. But for now, its travelling days are over. They were few, and relatively uneventful. No gyres, no pack ice. At the mercy of tides and currents, it was soon back on dry land; and now it lives on my mantelpiece, along with other strand-line strays and vagabonds.

Stella Maris

In his poem 'A Wreck', Michael Symmons Roberts writes of a mass
stranding of starfish:

> Countless five-toothed cogs mesh
> as a powerless machine. Dead
>
> and dying are irreconcilable.
> Once the ghost is given up, they dry
>
> to ornaments, inanimate as shells.
> In our vanishing bright future
>
> human hearts will be like these
> tight windmills with a carapace.

I'm standing on my own beach, surveying a similar scene after
a storm tide: thousands of 'dead and dying' on the sand. I pick my
way around and between them, as if touring a battlefield in the
dreadful aftermath, shocked by the scale of carnage, helpless in
the face of so much suffering. The wind is still keening, and loose
sand races over the surface like smoke.

On this occasion the casualties are all *Asterias rubens*, known as
the common starfish. It's the familiar orange or yellow type found

on all British shores, quite chunky, as big as my hand. Just as frequently found here is the sand star, which is smaller and flatter, with a dull brown colour that gives it camouflage and a row of tiny pale spines along the edges of its arms.

Starfish are well-loved creatures, found pictured in friendly attitudes in children's books and posters and on beach towels and clothing. There's Patrick Star, the protagonist's dim-witted but amiable sidekick from the animated television series *SpongeBob SquarePants*; and Peach, the lovable starfish character from the film *Finding Nemo*, available from the Disney Store as a soft plush toy with eyes and a smile.

The real lives of these creatures are more curious, and decidedly less cuddly. There are thought to be about two thousand species, including at least one – the crown of thorns starfish – which is venomous and dangerous to humans, but the true number is not known, and new discoveries are being made all the time. They are not fish, but marine invertebrates more closely related to sea urchins and sand dollars. In place of blood, they have a water vascular system. One species, the ochre starfish, has been observed pumping itself full of cool water to regulate its temperature when exposed to the sun at low tide. Some can switch gender. Some are aggressive predators, and can push their stomachs outside their bodies in order to smother and force open large prey such as clams and mussels. Others have a prodigious appetite for fleshy coral polyps which they liquefy with their digestive juices, each individual capable of turning coral reef to skeleton at a rate of six square metres a year.

But perhaps the most distinctive and well-known fact about the starfish – or 'sea star', as marine scientists would prefer us to call it – is that it can regenerate lost limbs.

It's not the only creature credited with this ability. Lizards can regrow their tails, spiders can grow new legs, and as every child knows, if you chop an earthworm in half you get two worms. Actually, this last example is a matter of debate, though one biologist, Professor G. E. Gates, devoted a lifetime's study to the earthworm, and concluded that it is 'theoretically possible' to grow two worms from one bisected specimen. (Professor Gates spent twenty years studying earthworm regeneration, but he published relatively few of his findings, because 'little interest was shown'. It must require great tenacity and strength of character to go on investigating something so specific in the face of the polite indifference of your peers.)

Humans are capable of regeneration, too. It's a staple of science fiction, from *Doctor Who* to *Friday the 13th*, but it's also a fact of life. Children who lose fingertips in accidents are able to regrow them, though without the fingerprint. The human liver, too, is well known for its powers of regeneration; it can regrow from as little as 25 per cent of its original tissue. It seems this was known by the Ancient Greeks, which might explain how Prometheus was able to go on surviving, in spite of being tortured by having Zeus's eagle feast on his liver day in, day out.

However amazing the regenerative powers of livers and fingertips, there are still more dramatic manifestations in the animal world, in species very far removed from our own human experience. Sea stars take it to extremes. Not only can a star regrow a lost arm, but for some species one arm alone is enough to regrow the whole animal.

We tend to be a bit squeamish about severed limbs, especially hands; they are both alive and not alive, which gives them a place, along with robots and possessed puppets, on Freud's list of the *unheimlich*. The horror genre has exploited this 'shudder factor' again and again, in films such as *The Beast With Five Fingers* and *The Crawling Hand*.

But for the sea star, it's simple: all the vital organs are housed in the arms. The arm *is* the animal. In some cases, all that's required for complete regeneration is just a small piece of a severed arm. The human equivalent might be to take a severed fingertip and grow a whole new child, which really does sound like the stuff of science fiction . . . although you could say we are taking tentative steps in that general direction, with work under way to use stem cells to grow genetically matched tissue, and ultimately whole organs, to replace those damaged by disease or accident.

Sea stars are often kept as aquarium specimens, and one fish-keeping forum has several threads about regeneration. 'I have a starfish and it has dropped one of its arms,' says a visitor. 'The arm it dropped is slowly moving around my tank. Does anyone know what is going on? Is the arm alive or just trying to freak me out?'

Another begins: 'Bought a red starfish a week ago, came down this morning and it was missing an arm. Will it grow back?' Later that day, the same correspondent sounds rather less relaxed: 'Came home from work to find the missing arm crawling up the glass, that's creepy! Will the arm die?'

The final instalment carries a note of hysteria: 'I went to check the progress and see if anything was left of the star, and the one arm that was left had crawled itself under a rock!. It's still alive!!!'

Who needs horror films when you have a saltwater aquarium?

*

The star has been a powerful human symbol since our earliest days. Ice Age star maps have been found in the painted caves at

Lascaux and El Castillo, and stylised stars appear on artefacts from practically every ancient civilisation. Whether with four, five, six or more points, the star has been pressed into service by a huge range of groups and causes. It appears on more than twenty national flags; it's an important religious symbol in the Islamic, Jewish, Baha'i and Christian faiths; and it's associated with Freemasonry, paganism, Satanism and witchcraft. To the Pythagoreans it represented an ideal of mathematical perfection; it has specific meanings in heraldry; and it has served as an emblem of both Communism and Fascism. It stands too, of course, for high ambition – we are frequently exhorted to 'reach for the stars', and those who have achieved stellar status in film, music and sport are worshipped from a great distance, just as the literal stars were by ancient civilisations. So diverse is the list of symbolic meanings that the star has become a profoundly enigmatic, almost an unreadable shape.

One of the ways in which Christianity has adopted the star is as a name for the Virgin Mary, mother of Christ. The ninth-century title *Stella Maris* emphasised Mary's role as a guiding star to travellers, and there are churches and schools – including one just up the coast in Liverpool – dedicated to Our Lady, Star of the Sea.

That beautiful title comes to mind when I see those other regulars on my beach: the Ophiuroids or brittle stars. They share with starfish a basic form and pentaradial symmetry, but in other ways they are quite differently made: leggy and delicately lovely, flexible and articulated.

I picked one up here a few months ago, and its central disc was decorated with an exquisite flower pattern, five pairs of joined petals like rosewood inlay. Others have banded arms and a tiny five-pointed star at the centre. Some discs are marked with a sort of beadwork; others are smooth and black as jet, with white spines like teeth on a comb.

Why these creatures should be quite so ornamental, I don't know; many of them spend their whole life in the dark, on the deep sea floor, using their supple arms to propel them along in a crawling motion, gathering particles of plankton and bacteria as food.

At least one species – *Amphipholis squamata* – has adapted to the darkness of its abyssal environment by acquiring the ability to create its own light, through bioluminescent signals given off from its arms. Bioluminescence is not very well understood, but it seems to serve multiple purposes, from deterring predators to attracting a mate. The ocean is alive with these natural lights, most of them in the deeps, where sunlight doesn't penetrate.

Others occur near the surface too. Some species of squid which live in shallower waters use bioluminescence as camouflage, to blend in with moonlight. And brilliant light can sometimes be seen sparkling on disturbed water; the light is emitted by algae called dinoflagellates as they are jostled; in the wake of ships, for instance. On a journey documented in his book *The Wild Places* (2007), Robert Macfarlane visits the westernmost tip of the Lleyn Peninsula in North Wales. On a moonless night, he goes for a swim, and finds himself surrounded by luminescence:

Where it was undisturbed, the water was still and black. But where it was stirred, it burned with light. Every movement I made provoked a brilliant swirl, and everywhere it lapped against a floating body it was struck into colour, so that the few boats moored in the bay were outlined with luminescence, gleaming off their wet sloped sides. Glancing back, the cove, the cliffs and the caves all appeared trimmed with light. I found that I could fling long streaks of fire from my fingertips, sorcerer-style, so I stood in the shallows for a few happy minutes, pretending to be Merlin, dispensing magic to right and left.

Certain kinds of bacteria are luminescent, too. These are almost certainly responsible for the mysterious light effect in tropical waters, known as 'milky seas', where an area of the ocean glows blue-white under the night sky. In the past this was the subject of folklore. It was encountered by the fictional adventurers in Jules Verne's classic 20,000 *Leagues under the Sea* (1870), when the Professor is able to offer his faithful assistant Conseil an explanation not a million miles away from the truth as we understand it today:

'But, sir,' said Conseil, 'can you tell me what causes such an effect? for I suppose the water is not really turned into milk.'

'No, my boy; and the whiteness which surprises you is caused only by the presence of myriads of infusoria, a sort of luminous little worm, gelatinous and without colour, of the thickness of a hair, and whose length is not more than seven-thousandths of an inch. These insects adhere to one another sometimes for several leagues.'

Conseil is in good company; there have been hundreds of sightings of the same phenomenon reported by mariners over the centuries, and now it's even been seen from space. Satellite imagery has shown one illuminated area stretching for over fifteen thousand square kilometres, off the Horn of Africa. A milky sea, it's now thought, is caused by an enormous congregation of luminescent bacteria, and lasts for just a few days at a time; but very little is understood about the conditions which make it possible.

*

The abyssal depths are still mysterious to us. D'Arcy Wentworth Thompson, author of the classic work of scientific literature *On Growth and Form* (1917), notes that 'the great depths of the sea differ from other habitations of the living, not least in their eternal quietude. The fishes which dwell therein are quaint and strange; their huge heads, prodigious jaws, and long tails and tentacles are, as it were, gross exaggerations of the common and conventional forms.' How have these exaggerated features developed, he asks, and to what purpose? Is natural selection the whole story, or are there other, more 'mechanical' forces at work to modify the forms of living things? He uses mathematics and physics to analyse questions of biological form, arguing that factors such as gravity and surface tension are crucial in shaping plants and animals.

As technology has enabled humans to start exploring the remote ocean depths, we have made surprising discoveries about the adaptations which make life possible here. One of the strangest has been the revelation that this is not a world of complete darkness. Parts of the deep sea are like the night sky. Constellations of brittle stars light up some of the world's gloomiest places, along with, amongst others, the angler fish, the splitfin flashlight fish, the sparkling enope squid, and numerous bioluminescent species of comb jelly, clam and sea slug.

The brittle star's luminescence is at its brightest when the individual is brooding; it's affected by other variables, too, such as seasonal changes and water temperature. But if a specimen is observed over a period of time it's possible to assess its state of health by how brightly it shines. Like starfish, brittle stars have a water vascular system. The water they inhabit is pumped directly into their bodies, and they have no way of filtering toxins out of it. This makes them highly susceptible to damage from pollution of all kinds; oil spills in particular have been known to devastate sea star populations. This vulnerability, combined with its special ability to generate light, is giving Amphipholis squamata a role as an indicator of water contamination. In one study, samples of sediment and seawater from different areas of San Diego Bay, an area which is suffering serious metal pollution, were collected and put into tanks with brittle stars. The bioluminescence levels were then observed and compared. Those stars in water from the mouth of the bay, where the toxins are more concentrated, gradually lost their brightness over the six-week period.

There's a great deal we don't know about these creatures. Amphipholis squamata is polychromatic: it comes in at least six colour varieties. One looks like a silver coin with slender, blue-white limbs, toothed like those plastic ratchet straps you sometimes find on packaging. Another is yellow, with a design in the centre like the pattern of pips you see if you make a transverse cut through the middle of an apple. There's a beige variety and a black one. Some biologists think that they might all be 'sibling species': very close but genetically different. No one's sure.

Brittle stars are still essentially mysterious. But we know they're highly sensitive to pollution. It's not difficult to imagine that in some of our ruined waters the dazzling show could gutter and fade, and one by one the stars go out in the deep undersea places.

Squirt

For a second I think it's a human ear. Body parts are washed up from time to time, and it's occurred to me that one day I might make some gruesome discovery, given how often and how thoroughly I explore here.

Murderers often underestimate the strength and complexity of currents and tidal systems, a failure which has sabotaged many an otherwise 'perfect' crime. Of course, bits of bodies on beaches are statistically more likely to signify suicide, accidental death, or erosion of old burial sites during storms. Arms or legs sometimes drift great distances, if they're well wrapped in sturdy clothing. They are occasionally identified as belonging to victims of plane crashes hundreds or even thousands of miles out to sea. Recently there has been an unexplained spate of feet, clad in trainers, washed up on the shores of the Salish Sea in British Columbia. A number of theories have been advanced: that the feet belong to people who jumped from river bridges, or to Mafia victims weighted down and dumped at sea, or to casualties of the Asian tsunami in 2004. Certainly it is known that the feet easily become detached from a submerged corpse, and that if they are still enclosed in socks and strong shoes they can stay more or less intact for years.

But an ear?

There's really quite a likeness: it's pink, fleshy, just the right size. But when I poke it experimentally with a gloved finger I see, with a mixture of relief and disappointment, that it's not an ear. It's not any part of a person. So what is it? Now I've turned it over I can see it has two little raised openings at one end, and that's what gives it away: it must be a sea squirt.

In the glossary to Under the Sea Wind (1941), Rachel Carson writes: 'Sea squirts have leathery, saclike bodies, and when touched eject spurts of water from two openings like short teakettle spouts. They grow attached to stones, seaweeds, wharf piles, and the like, straining food animals out of the water by passing it through an elaborate system of internal structures.' Her summary may account for why I've never seen one on this beach before: there are no stones and very little seaweed growing in this area, and few underwater structures, though a few miles north there is the second-longest pier in England, which I guess might be home to some of these unfamiliar creatures.

What makes beach finds special is that they arrive here from so many different sources and directions: from the hands of walkers and picnickers; from the air; from underneath the sand; and of course from the sea. In an age when science has unlocked so many of Earth's secrets, and almost the entire planet has been mapped and imaged, our oceans and shores remain relatively unexplored. Each new discovery presents questions and mysteries.

Sometimes these questions are of genuinely global significance. The evolutionary biologist Jonathan Eisen, who has spent the last eight years dredging up and analysing samples of seawater from around the world, has found gene sequences that are radically different from anything seen before and that appear not to belong in any of the three domains in which we currently categorise living things. Where it comes from and what the implications are for our understanding of life on Earth, no one knows yet.

Other shoreline discoveries are on a smaller scale, but they can be just as engrossing, for beachcombers if not for biologists. Sea squirts are familiar enough to science, but this is my first encounter with one. There are thought to be at least two thousand species, most of them filter feeders, although recently a large, predatory type two feet tall was discovered anchored to the deep sea floor off Tasmania. Even without this startling kind of exception, they are a disparate group, varying from a couple of millimetres to a metre in length and taking a wild variety of shapes, forms and colours. They're also known as 'tunicates', because of the tough outer test or tunic which covers the body; it's not a shell, but is made of living tissue.

I can't make a positive identification of the one I've found – it looks quite like *Ascidiella aspersa*, but it could be *Distomus variolosus*, or some other again. The only thing I'm fairly confident about is that it's not a 'solitary' species, but one of the type which clump together in large communities.

The life history of the sea squirt is unique. It begins life as a free-swimming creature, similar in appearance to a tadpole. At this stage it has a nerve tube and a primitive backbone or notochord. After a few days in this tadpole state, the squirt finds some kind of surface, depending on what's available locally. Mine may well have found its way to the pilings of the pier. It then latched on, joining others of the same species which were already attached there. Together they formed a colony which looked like a bloom of pink flowers; each petal a separate individual.

From the moment of latching on, the squirt relinquishes its freedom to move and becomes sessile, or fixed. Having found a place to settle, it stays there for life. It now metamorphoses, absorbing the notochord and recycling the nerve tube into a more sophisticated cerebral ganglion. In the past, this metamorphosis has been a source of serious confusion; at one time it was suggested by some natural historians that the sea squirt started life as an animal, and then became a plant. More recently, wilful misunderstanding about the exact nature of the nerve tube gave rise to an in-joke amongst scientists that sea squirts, like senior academic colleagues, have a habit of 'settling down and eating their own brains'.

The moment of attachment has its own secrets, which are gradually being unpicked by medical scientists studying organ transplantation. For patients who need a replacement heart or liver or bone marrow, there is an ever-present danger that the body will reject the organ and the procedure will fail. This is a threatening cloud on the horizon for anyone receiving a transplant. The reasons behind acceptance or rejection are still largely mysterious, but researchers at the University of California, Santa Barbara see the behaviour of sea squirts as a possible analogy that might help them understand the problem better.

They hit on the idea of looking at what happens when a type of sea squirt known as the star ascidian identifies a potential attachment site and lands there next to another of its own kind. The squirt ascertains the presence of a suitable surface by using its sense of light and touch. It then puts out finger-like projections called ampullae, which make contact with the neighbouring individual and recognise it either as 'self' or 'non-self'. If the two are related, they fuse; if not, they reject one another. Current research focuses on how the signals received on the surface of the ampullae are translated inside the circuitry of the cell, where the final decision about acceptance or rejection is made.

The specimen I'm holding looks like a simple creature, and a harmless one. But sea squirts are causing ripples of anxiety as well as excitement in the scientific community. Over the past century, some species have travelled in huge numbers on the hulls of ships, and have established themselves in harbours all around the world. Colonies can grow very quickly, and in some places they're smothering populations of native animals.

On the other hand, they are a welcome food source for fish, birds, worms, starfish and molluscs; and in other parts of Europe they play a part in the human diet. They often sneak onto the menu under the general heading of 'seafood', but the French *violet* and *figue de mer*, and the Italian *limone di mare* and *uovo di mare* are all types of sea squirt. While some people have dismissed them as bitter or pungent, the cookery writer Jill Dupleix is a fan. She describes them as 'funny looking round, knobbly, shells . . . Inside the flesh is a bilious yellow and tastes like a sweet raw mussel once you have winkled it out.' But she concedes that 'you have to be brave to eat something so yellow'.

They seem to be entirely absent from the British culinary tradition, which is a shame because they sound like a real flavour experience and I'd like to try one. Perhaps not this one, though. It's not the lurid colour, nor the rubbery and rather granular texture; it's that momentary resemblance to a human ear that has taken the edge off my appetite. I think I'll wait till next time I'm in Amalfi or Marseilles.

Queen's Cup

The sea is a notoriously powerful destructive force. Many of the things washed up on beaches are proof of it: they lie scoured, mangled and torn; destroyed or transfigured by the violent journeys they've made. A high sea can break a boat in pieces, smash sea defences, drag a child off a harbour wall.

How miraculous, then, to find something as delicate as a china cup here. I'm walking on the beach, the morning after a storm has stretched the sand smooth as a tablecloth, and here in the placid sunshine is a cup, resting modestly on its side, as though it has been knocked over by a careless elbow when someone reached across to pour the tea.

It's white, with an old-fashioned design around the edge. There's a chip with a hairline crack running diagonally down the side, and a small patch of discoloration. The inside is lightly encrusted with sand, like coffee grounds. When I turn it over I find the Ridgway Potteries mark underneath, and an emblem, only very slightly rubbed away. It shows a lion wearing a crown and bearing a globe in its paws, along with the legend: THE CUNARD STEAMSHIP COMPANY LIMITED.

Cautiously I look up and down the beach. I'm in the habit of picking up odd bits and pieces and taking them home. But this is different. This is a desirable object. I feel like a shoplifter.

A minute later a man approaches, carrying several heavy plastic bags, one of them dripping a brownish liquid. He stops to chat about the wreck of the Victorian cargo ship the *Star of Hope*, which is lying exposed in a watery dip a bit further down the beach. He shakes his head lugubriously. 'I don't think she'll sail again,' he says with devastating understatement. He's been coming to see her for fifty years, hoping to find her 'out'. His uncle first brought him here as a boy in 1942, and did I know that for a time during the war the beach was planted with wooden stakes, each one tall as a man, as a defensive precaution to stop enemy aircraft from landing? I slip my precious find into my jacket pocket, impatient to get it home and examine it properly. He talks on, until I ask what's in the bags. 'Just a few shells,' he mutters defensively, and takes his leave.

If I'd found this cup fifteen years ago, finding out about its history would have been quite a challenge. No doubt there would have been a handful of collectors specialising in Cunard tableware, but how would I ever have found them? Now I have only to tap a few words into Google and there they are. The internet is a window onto the inexhaustible variety of human interests and obsessions. However singular, however preternatural the subject, there is a place where people gather to discuss its arcana. Often, tracking these enthusiasts to their digital clubhouses, I think of the words of Louis MacNeice: 'World is crazier and more of it than we think'.

A search yields two or three genuinely relevant results, and within seconds I've found a website run by a Cunard expert called Richard Villa. Richard is not an antiques dealer but a personal collector, and his website features photographs and descriptions of the many different designs of soup bowl, salad plate, egg hoop, consommé cup, hors d'oeuvre dish, teapot, creamer, salt dip, sherbet cup, demitasse, fruit saucer and oyster dish used on the great transatlantic liners since they first set sail in the mid-nineteenth century.

It's taken Richard a lifetime to build his collection. His fascination with the Cunard Line originated in his student days, when he had a part-time job as a tour guide aboard the Queen Mary. By then she had been 'retired' and taken to Richard's home town of Long Beach, California, to be moored and used as a tourist attraction. He started buying the odd item of paper memorabilia from the ship, and then furniture too. Bitten by the Cunard bug, he became a self-confessed cruiseaholic and has sailed on thirty-six cruises so far, many of them on old retired liners since dismantled and scrapped.

But his china collecting didn't take off until the arrival of eBay, when he started to sell duplicates from his collection of ship memorabilia, and in doing so found items of china from the Queen Mary for sale. Suddenly he realised he had always wanted to collect Cunard china. To begin with, his modest ambition was to acquire four place settings of Maddock Ivory Ware, the sort used in the first-class dining room. He now has 125 pieces and counting. On his website there is a series of period photographs of Cunard dining rooms and kitchens, showing the china in use in its original context. Then there's one of Richard himself, sitting in his own home at a table set with his historic collection.

Naturally I'm hoping that my cup might have some rarity value. For a while, I'm excited about its possible provenance on the Mauretania, or even her doomed sister ship the Lusitania, which met with disaster not far from here, off the Irish coast. She was torpedoed by a German U-boat on 7 May 1915, and sank in eighteen minutes flat, killing over a thousand passengers and crew. The bodies of the dead are said to have floated all over the Irish Sea for days afterwards, and fishermen were paid a reward for each one they recovered. Could my cup have been lost in this disaster, and have been in the water ever since?

The pattern around the edge is easily identified as Greek key. But then it becomes more complicated. China of this design existed in a number of variants: with and without a gold band on the rim; with four or five 'breaks' in the pattern; and even a rare and enigmatic version intertwined with flowers and vines. The detail of the design can tell you which ship and which class of dining room it belonged to.

In the course of his buying and selling, Richard found confusion and contradiction over the provenance of some of the china, and was able to gain access to archives to carry out research and create an authoritative list of pieces and their histories. On his website I find a clear photograph of a cup identical to mine, but in mint condition, and complete with saucer. The note underneath reads: Cup – 3⅝" dia. x 2½" high. Used in Third Class aboard the Queens.

Not the Lusitania, then, but one of the two great flagships, the Queen Mary and Queen Elizabeth. And not the glitter and grandeur of a first-class dining room, but the simpler environment of third class, where the less well-heeled passengers took their meals.

I turn the cup in my hands, looking again at its faded design, its chipped edge. I want to understand the life it had before it came to me. The story of Cunard is intimately connected with the story of Liverpool, and that's where I need to go to find out more.

★

The records of the Mersey Docks and Harbour Company are held at the Maritime Museum on Liverpool's Albert Dock. This dockland development felt bleak and empty in its early days, when we would go there to take our children to the Tate Gallery. They loved the Tate. Their two highlights were a piece of contemporary sculpture – a silver tap with drops of silver water suspended underneath – and the 'nappy macerator' in the baby-change room, which made a noise like a bear eating bricks. Between them, these features pretty much guaranteed a good day out.

In those days it wasn't easy to buy a sandwich round here. There was a tourist shop selling Liverpool scarves and Fab Four keyrings; a poster gallery, and the unavoidable jellybean shop, whose colours and scents were part of the very fabric of my children's dreams, a place where a small bag of sweets would set you back three quid even then. There always seemed to be a cold wind swirling discouragingly around the dock, and the floating map of Britain on which the Granada TV weatherman delivered his daily forecast bobbed on the clouded and sullen water. It was as if the whole of the old docklands was in open revolt against this attempt at reclamation and gentrification. It was an island of posh, out on a limb, cut off from the city by the A5036.

Things have changed. The route you walk from the railway station to the waterfront now takes you through Liverpool One, the flagship brand for the £900 million retail-led regeneration of Liverpool city centre. A shopping precinct, to you and me – but an impressive one, I have to admit. It's at the heart of the Paradise Project, a name redolent of hubristic over-ambition but actually taken from a local street name. There's a Paradise Street in just about every city, usually in one of its least favoured districts, a testament to failed optimism. However, on this day at least, Liverpool One seems to be a recession-free zone: it's thronged with shoppers swinging glossy carrier bags.

Down at the Albert Dock things are different too. The place

looks plumper and more polished, and new buildings are going up all along the waterfront, including what looks alarmingly like a multi-storey car park shrouded in scaffolding, rising to block out the view of the iconic Liver Building.

In dry dock opposite the Maritime Museum are its two largest exhibits: a pilot boat and a three-masted schooner. The building itself houses the International Slavery Museum and a number of exhibitions, including one devoted to a trio of shipping tragedies: the Titanic, the Empress of Ireland and the Lusitania. They say disasters come in threes, and the loss of these three ships, all between 1912 and 1915, must have had a devastating effect on their home port of Liverpool. Perhaps they signalled the end of the great Edwardian boom and the start of a long, slow decline in the city's fortunes.

Eighty-five years on, the Lusitania is still commemorated on both sides of the Atlantic. Here in Liverpool, an enormous four-bladed propeller, one of three salvaged from the wreck, is displayed on the quay outside the museum, and every year on 7 May a memorial service and laying of flowers is held there. In earlier years, quite a number of survivors of the disaster would attend and take part in the ceremony, but as the years went by they were fewer and fewer in number. The last survivor, Audrey Lawson-Johnston, a babe in arms at the time of the sinking, died in January 2011.

Inside the museum, I follow signs to the Maritime Archives and Library. It's a gem of a place, tucked away on the second floor: a little world of its own with its treasure-store of books, charts and microfiche records. Here is all the information I could possibly want on the Queens Mary and Elizabeth.

Both were built in the 1930s, in a burgeoning spirit of patriotism, and in response to Germany's ambitious shipbuilding programme. We couldn't have the Germans humiliating us with bigger, faster and smarter ships. The Queens were the British retort, and they were very big, very fast and very smart indeed, at least to begin with.

Both had chequered careers, partly because of the unlucky timing of their construction. The varnish on their decks hardly had time to dry before war broke out, and they were commandeered and pressed into service as troopships. They were painted grey, stripped of their luxurious furnishings and fitted with bunks, hammocks and guns. During this military phase, the Queen Mary suffered a catastrophic collision with the anti-aircraft cruiser HMS Curacoa, slicing it in half and killing three hundred men. When the war ended, the two ships were handed back to Cunard and restored to passenger service, but the glory days of the transatlantic liners were numbered, and both were 'retired' in the 1960s. Mary lives on at Long Beach; Elizabeth was taken to Hong Kong harbour to be refitted as a university, but was destroyed in a suspicious fire in 1972.

I look up the precise physical details of the two Queens here in the library, which houses the complete set of Lloyd's Register of Shipping. This amazingly detailed record of the growth of the maritime industry runs from the 1770s to the present day, and takes up two complete walls of shelving, floor to ceiling. Some of the later volumes are so large and heavy that there are instructions posted on the shelf edge, telling readers how to lift them down without damaging them. You could certainly break your foot if you dropped one on it.

Lloyd's Register 1935–36 lists all the important details of every vessel built that year. Here's the Queen Mary, built by J. Brown & Co. Ltd, Clydebank, owned by Cunard White Star Ltd and registered at the port of Liverpool. She weighed 81,235 tons – an absolute monster compared with all the other ships listed in these pages. She was powered by 16 steam turbines and 24 water tube boilers.

I track the *Queen Elizabeth* down to the 1941–42 volume, and find that she was even larger than her sister, and capable of carrying 2,283 passengers and over a thousand crew. By virtue of its wartime publication date, this volume has an exciting extra feature: the cover is stamped with the words 'SECRET – see notice inside cover'. Inside, a label has been pasted which reads: 'The book and its supplements are invariably to be kept locked up in the safest place available when not in use. Every effort must be taken to prevent them falling into the hands of enemy agents.'

I cast a furtive glance around the room. The librarian is helping two elderly American sisters operate the microfiche reader. A man is sitting reading a book called *The Making of Liverpool*. A woman opposite, surrounded by tottering piles of books, makes notes furiously on a page of neatly ruled columns. She looks up, catches my eye and starts to talk. She tells me she's trying to find her father. The family always said he was a local man, went to sea and never came back, missing presumed drowned. But now she's been diagnosed with a rare genetic disorder that only occurs in people of Asian descent, so where does that come from and what can it all mean?

I can only imagine what it's like to live with a mystery like this, to be so uncertain of your parentage. Those of us who have known our parents and our grandparents, who have looked at photographs with them and heard them talk about their lives, are the lucky ones. We can look back two or three generations, maybe more, and trace a clear, unbroken line. We know where we've come from. Presumably if the woman opposite had not fallen ill she would never have questioned her family's official line.

Then there are the American sisters, who are spending a few days in Liverpool trying to piece together the story of their father's death; they have the name of the destination but not the name of the ship that came to grief, nor the exact year. The librarian is endlessly patient and painstaking. I'm beginning to see that enquiries like these are the greater part of what goes on in here. It's where people come to search for the truth, often rewriting family history in the process. 'We had no idea . . .' says one of the sisters in a bewildered tone. 'We were led to believe . . .' They repeat these phrases again and again as the patient librarian turns the wheel on the microfiche reader, and the miniature evidence, hidden away for so long in a box in the drawer of a filing cabinet, springs up bold and legible and brooking no argument.

Surrounded by all this intensity, my own purpose here looks strangely abstract. It's not my family and my personal identity I want to explore. It's the object I have wrapped thickly in newspaper in the rucksack at my feet, and other objects like it: the things people have taken to sea with them, and lost there.

*

The Cunard era brought work and prosperity to Liverpool and the surrounding area. It linked the city with New York, a connection which remains live today. The merchant seamen who staffed the ships in the 1940s and '50s, known as the 'Cunard Yanks', brought back the latest music, fashion and consumer goods, and have been credited with bringing the sound of rock and roll to Britain for the first time. The men had split lives, shuttling between the cramped, blitzed streets of their home town and the dazzling glamour of New York. They worked very hard and their wages were not high – £27 a month for a steward in 1952 – but it was all about adventure.

One of the men has described the experience as 'going from a black and grey world into one of Technicolor'.

Even for those left behind, the ships brought a touch of exoticism to post-war Liverpool, and smaller towns along the coast. You could stand on these beaches and watch them nudging out of the estuary and cutting a road into the open sea. After dark, they were lit like floating towns. On a summer night you might catch a scrap of dance music on the breeze, like a wisp of scent.

There must have been tokens of those romantic journeys lost overboard and washed up here – a silk scarf, a cigarette packet, a photograph, a ticket. To whoever found them as they walked on the shore, these things would have spoken eloquently of travel and opportunity, of another, less ordinary way of life.

*

I leave the archives and wander round the museum. The words of Joseph Conrad are displayed, huge and portentous, on the gallery wall: *As long as men will travel on the water, the sea gods will take their toll.*

If only my teacup had a gold band around the rim, it might belong here, in a display of artefacts from the *Lusitania*. There's a deckchair, pulled from the sea near Kinsale by a young fisherman called Patrick O'Driscoll, the day after the sinking. A photograph taken many years later shows Patrick as an old man, posing with family members in his garden, the deckchair taking pride of place at the centre of the picture. And a sofa cushion, found by Royal Navy Seaman Henry Grew, during his search for bodies in the Irish Sea. You can see the tear he made in the fabric as he hauled it out of the water with his boathook.

These things speak intimately of the lives being led on board the Lusitania in the days and hours before disaster struck. Even as the order was given to fire the torpedo, someone was sitting in this deckchair looking at the view from the deck, or dozing on this cushion after a good meal. It may have been wartime, but there was still frivolity and luxury, at least for the privileged few. Spread at the back of the display case is a white paper fan, perfect and intact, one of those given to the ladies at the Captain's Gala Night in the ship's ballroom.

A little further along the gallery is a lifebuoy apparently made of leather, cracked and weathered like an old face which has seen it all. This was taken as a souvenir by the skipper of a fishing boat from Kinsale who rescued some of the survivors. Pulling these shivering and traumatised people from the sea and bringing them to safety must have been an unforgettable experience. He was just doing what he could to help in appalling circumstances, but couldn't have foreseen the global implications of the tragedy: the surge of public anger on both sides of the Atlantic, and the role it would play in influencing the eventual decision of the United States government to enter the war two years later.

In the days that followed the sinking, there were ugly consequences here in Liverpool, with mobs smashing German shops and attacking the homes of neighbours with German-sounding surnames. But there's no sign of vengeful fury in the faces of the people pictured waiting for news outside the Cunard Line offices in Water Street. Whilst the passengers were predominantly American, most of the crew were Liverpudlians, and the personal losses on both continents must have been terribly hard to bear.

Clips from British Pathé newsreels show the street packed with people, all with these same unreadable faces. A young woman – no more than a girl, really – stands alone in the crowd and jiggles a crying baby in her arms. Then she turns and gazes straight through the camera, her face blank, expressionless. No sound, of course, but in any case you can see that hardly anyone is speaking. There's an icy tension about the way they are all standing and waiting in silence. Now the film cuts to Lime Street station, where survivors and relatives are leaving, either reunited or alone. Here there are a few gestures of distress: one woman pulls out an enormous white handkerchief and wipes her eyes; another places her hand on her companion's arm for support. The faces register shock, but we don't see anyone wailing, collapsing, embracing. Were people more stoical in their grief back then, or did they prefer to mourn in the privacy of their homes? Or is it simply that, one year into the war, a sense of resignation, a kind of numbness, had already set in?

*

I can only imagine who might have tossed this cup overboard. It could have been a mischievous child, a lovesick student, a honeymooning couple having their first row. Was it dashed over in anger, despair or exhilaration?

Was it done for a dare, or in a spirit of curiosity and fun, in just the same way as everyone likes to throw a pebble into a lake or drop a twig from a riverbank and watch it carried into the distance? Messages in glass bottles have been known to survive for hundreds of years; the impulse to make a connection with someone we have never met is a universal one, and whoever threw this cup overboard half a century ago may have given a thought to whether it might be found, some day.

Archive stock sheets from the early 1960s, towards the end of the Cunard era, hint at the working lives of the crew on board the luxury liners. Like the cabin crew on aeroplanes today, they worked in confined spaces and under testing and unpredictable conditions, for a clientele who had high expectations of service. They were also held responsible for looking after equipment of various kinds. The sheets list the number of each item of 'chinaware, earthenware, glassware etc.' on the *Queen Mary*, along with the value of each in pounds, shillings and pence. The documents show how scrupulously staff were required to account for their stock. The task of auditing alone must have been stressful, boring and arduous; every item of tableware and kitchenware would have to be fetched out of the lockers, its condition checked, the numbers counted and entered by hand on the sheet, and the sums – those pre-decimal calculations that look impossibly difficult now – done without the aid of a calculator. You can bet there was pressure to keep losses down; running a cruise line was a competitive business, and if too much was wasted, there would be unwelcome consequences.

I can speculate that my cup – perhaps after all a 'breakfast cup' rather than a teacup – may be one of the 4,166 listed under 'breakages or loss' in the course of Voyages 405–427 (inclusive). That's an average of 181 cups per voyage. How many of them ended up thrown overboard? I wonder. What makes my own find so mysterious is the fact that it has survived, to be washed up more or less intact at least fifty years later. I email Richard Villa through his website to ask him what he makes of this. He writes back immediately: he's amazed, even though Cunard china is very strong. He uses his first-class pieces daily, and puts them in the dishwasher. It was, after all, hotelware, well made and designed for heavy use.

Ridgway's, the company which made it, was in its heyday one of the most important potteries in Staffordshire. It was based first in Shelton and then in Hanley, the 'capital' of the Potteries: an

Industrial Revolution 'boom town' which features as 'Hanbridge' in Arnold Bennett's *Anna of the Five Towns* (1902). In the novel, the towns are collectively described as 'mean and forbidding of aspect – sombre, hard-featured, uncouth; and the vaporous poison of their ovens and chimneys has soiled and shrivelled the surrounding country till there is no village lane within a league but what offers a gaunt and ludicrous travesty of rural charms'.

The Potteries are about as distant as it's possible to be in England from the sea. No glamorous ships called here, and no keepsakes of their journeys washed up on its pavements or factory yards. Anna's life is one of suffocating provincialism. But manufacture and trade can open windows in the narrowest places. In the 1820s, one of the Ridgway company's most lucrative ventures was decorative pottery for the American market: plates, jugs and vases showing American buildings, scenery, the arms of the United States, portraits of Washington and other symbols of the New World. This early transatlantic enterprise foreshadows the Cunard connection over a century later.

★

Where has my cup been all those years? Was it buried in the intertidal mud, and uncovered by the action of the waves? If shipwrecks can come and go like this, then smaller objects can too; though a china cup – even a hard-wearing one like this – is a relatively fragile thing which surely wouldn't bear the weight of much sand.

Or did it spend those long years in the water? The sea has a way of holding on to things. It swallows whatever it's offered, without discrimination, and can keep it for decades, even centuries, drawing it into its own private system of currents which are still as mysterious to us as the corridors, spiral staircases, attics and cellars of the houses we visit in our dreams. At some moment the object is thrown onto a beach, perhaps many miles from where it was first snatched. Spatial and temporal distances have lost their meaning. So a Roman coin can lie on the sand in a tangle of bits of coloured plastic; in the world of the strandline, all things are equal.

I'll never know. But now it sits on my mantelpiece: my equivalent of the expensive antique vase or candlestick. It may be chipped, its design may be rubbed faint in places, and it still contains a few grains of sand; but I value these imperfections. They're like lines on a face. They're the marks of experience, describing in their own coded language the detail of where it was found, and the long journey it had made to get there. It's a thing of stories, and a survivor.

Winter

Deep Freeze

December 2010 is officially the coldest in a hundred and twenty years. It snows hard, and the snow sticks. On the night of Saturday 18 December, the Met Office station a few miles away in Crosby records its lowest ever temperature: minus 17C, an exceptional figure for a coastal location. The cold brings 'sea smoke' drifting over the water, reducing visibility to a few metres.

The following day, it's still well below zero. I set off to walk to the beach, closing the front door gently behind me to avoid breaking the portcullis of icicles hanging from the guttering above. This may be a fool's errand: roads are impassable and the railway line is snowed over. No one else seems to be going anywhere. But I want to see what the shore is like in these extreme conditions.

On an ordinary day, my route through the pinewoods and sand dunes takes a brisk half-hour. Close to dusk, after three hours' wading through snow, I stagger out of the dunes and straight onto a Rothko canvas: one of his troubled late pieces, when he had all but abandoned colour. The bottom three-quarters of it is an uninterrupted block of white. Above that, a streak of steel water. Then a broader strip of charred sky.

The world has collapsed to two dimensions. It's entirely empty, even of seabirds; stripped of colour, movement and sound. I'm determined to get to the sea, but the snow is a foot deep and progress is slow. I'm tired, and the cold is stifling. All I can hear is the creak of snow under my boots, and the splintery sound of my own breath.

*

A heavy snowfall which sticks, even on a sandy beach, is an unusual event on the English coast.

Ice is less rare. I have described in a poem the sheet ice which forms here, perhaps once in an average winter. It's 'one single sheet of sprung light', extending over huge areas of sand, appearing to make tentative links between distant points, to act as connective tissue between the estuary and the pier thirty miles away.

But ice is a surprisingly varied phenomenon, and the single sheet is just one of its manifestations. I've seen many others. A tideline made of frozen froth, stretching away into the indefinite distance, arranged in broad scallops and swags like a length of silk decorating a wedding marquee. Little heaps of 'frazil ice' in delicate shapes, like handfuls of sharp bones and feathers. Larger constructions like gleaming white bonfires ready to be lit.

Sometimes you get a mass of tiny icebergs and floes, thawing gently in the sunshine and leaking back towards the sea. They're a miniature reminder of the dramatic sight that greeted me on holiday in south-east Iceland, when I drove over the bridge at Jökulsárlón, where the glacier breaks up and blue icebergs float out to rest on the black volcanic beach.

There are days, too, when the sand is sharply textured, each hollow holding a lens of ice, and each ridge smeared with frost; every line, every shadow drawn in keen detail. As Robin Robertson observes, in his poem 'The Park Drunk':

What the snow has furred
to silence, uniformity,
frost amplifies, makes singular:
giving every form a sound,
an edge, as if
frost wants to know what
snow tries to forget.

It's always a thrill to see ice on the beach. But I've never seen anything quite like the legendary winter of 1963, when, as the locals still remember, the sea froze for a mile out from shore at Herne Bay in Kent.

This hardly ever happens in British waters. It rarely stays cold enough for long enough. When water freezes, molecular activity slows down and the molecules become fixed, forming crystals. But the presence of salt particles gets in the way of this process, unless the temperature is significantly lower – typically minus 1.8C. At Herne Bay, the Big Freeze went on for three months, and the consistently low temperatures allowed the ice to form in spite of the salt.

The climate is changing, and in unpredictable ways. Perhaps the day will come when I see the same thing here. On a different timescale, it will certainly come to pass. And when it does, it will be a return to something very familiar; after all, the area where I'm standing was once under ice three miles deep.

But for now, the waves go on moving and the sea remains liquid. It's the shallow water caught in pools or left lying on the sand as the tide recedes that gets caught, and transfigured, by a steep plummet in temperature in the early hours.

★

The snow thins and frays, and the final few metres of sand are exposed. The silence melts a little here too, enough for me to hear the muffled sighing of the waves. They seem to labour more slowly and heavily than usual.

When I turn and look back the way I've come – the way I must take to get home – the dunes are alpine, touched here and there with the last pale sunlight. Out to sea, the canvas is getting darker, losing definition. It's not yet four o'clock, but we're two days from the winter solstice; night is already pressing in, gripping the sky and the water.

SOS to the World

There can't be a beachcomber in the land who hasn't poked about speculatively in the strandline, hoping to find a message in a bottle.

They are beguiling objects. They speak to us of a deep human paradox: our need to be alone, and our need to connect with others. How are we ever to reconcile those two contradictory desires? Through the messy, complicated business of conversation? The tired old bureaucratic means of letter or email? That would take work. But the random find of a message in a bottle is a fresh and thrilling new idea. A total stranger reaching out and finding us. This could be the chance in a million which changes everything. A world in which there are messages in bottles is a world still trailing, in spite of everything, a few tattered rags of romance.

When I was nine or ten, my friend Julie and I had a dead letter box. It was a hole in a tree, in a field where we were not supposed to go. There was a warning sign nailed on the gate, and the farmer was said to be in the habit of pointing his shotgun at trespassers. Furthermore, to get to the tree you had to pass a haunted barn, a dilapidated structure made of corrugated iron. I used to hold my breath as I walked past it, a tried-and-tested method of warding off bad magic.

It was worth it for the thrill of receiving a letter. I still recall vividly the sensation of reaching up into the scratchy hole in the bark, scrabbling around blindly with my fingertips, touching moss and old leaves and owl pellets, and finding a wad of damp paper. Our letters were always neatly sealed inside envelopes handmade out of pages torn surreptitiously from school exercise books (though this in itself was a risky activity, since our teacher Mr Knifton was meticulous in checking our books, and had a kind of genius for spotting the telltale threads of a torn-out page still attached to the staples).

From an adult perspective, it looks like a whole lot of unnecessary effort. We could easily have passed each other letters at school, or walked round and put them through each other's doors. What was so secret, anyway, that it had to be written and sealed, rather than said out loud in the playground? I have no idea. I remember nothing about the content, only the method of delivery. Climbing that gate, skirting the field and retrieving the letter: these made for some of the most intoxicating moments of my childhood.

How much more exciting it would be to receive a letter posted in the mother of all dead letter boxes: the sea.

*

On the day I find my first ever message in a bottle, I also find my second. In fact, I find them within ten minutes of each other.

They're lying near the high tide line, fifty yards apart. I've been kicking through the seaweed and driftwood, turning up an especially diverse rubble of objects, including a chimneypot, a plastic can marked with a skull and crossbones, a blue and yellow sign saying *WARNING CONSTRUCTION SITE KEEP OUT*, a one-legged teddy bear and a whole cabbage.

I've also been looking at bottles – they are so many, and so various. Mostly whole, but here's the neck of one, snapped so cleanly it's hardly sharp at all, with threads of seaweed spilling out.

I'm thinking how exciting it would be to find a message inside one of these bottles. And then I do. I find one. It's unmistakable. The bottle is plastic rather than the more traditional glass, but I can see the roll of white paper inside.

It's a moment in which I question my own sanity. I thought of something, and it materialised. Am I turning into one of those people who go on about the power of visualisation? Am I henceforth, like Noel Edmonds, a believer in Cosmic Ordering? Will I, like Noel, be visited by two melon-sized orbs, which appear over my shoulders and represent the spirits of dead relatives?

Dismissing these anxieties, I pick up the bottle and try to unscrew the cap, but it seems to be superglued shut. I'll have to take it home and break into it there.

Fifty yards further on, with the exciting bottle stashed in my rucksack, I find the second. And actually my high spirits sink a little. The bottles are superficially different, but I can see straight away that the contents are identical: a roll of white lined paper, tied neatly with black thread.

This starts to feel less like a magical find, and more like something orchestrated. A treasure hunt, maybe? A game of some sort? Like a class of schoolchildren releasing balloons with messages attached.

Even so, I stash the second and head for home, where I stand the two bottles on the kitchen table, and sit staring at them for a long time. I think about what might be inside. I ask myself what I would like it to be.

I brace myself for disappointment; in case the paper is blank, for instance.

Then I open them. The glue is so solidly applied, I resort to sawing them open with a bread knife.

A little scroll of paper falls out of each.

When I undo them, and smooth them flat, the messages are identical. Each consists of two sheets of lined paper from a reporter's notebook. The notebook must have had its spiral binding removed, because the perforations along the top of each sheet are unbroken and intact. Even Mr Knifton would have been outfoxed.

The message, in each bottle, is a list. It's written entirely in capital letters, very neat and precise. Some words and phrases are underlined thickly in red. The thing that characterises the handwriting is its extreme precision, and the two copies are so alike that when I overlay them and hold them to the light there's almost no variation. The loops on the letters P, R and B are not completely closed, but their openness is uniform.

CONCORDESKI CRASH   PARIS

31/12/68	CONCORDESKI TU144
31/12/99	PUTIN PRESIDENT OF RUSSIA
25/7/09	BLERIOT FRANCE→ENGLAND
25/7/59	HOVERCRAFT ENGLAND→FRANCE
25/7/00	CONCORDE CRASH PARIS
1/2/85	FLITEFORM AIM GRP HEATHROW
	STEADMAN STARNES BATES
	McBOON NOV 89
1/2/05	WM GRAY HEYSHAM LA3

It reads almost like a minimalistic experimental poem: the gaps and ellipses, the odd connections and juxtapositions. It certainly has something of the provocation of a poem which withholds more than it gives away.

The obvious theme is aviation, from the first successful flight across the English Channel by Louis Blériot in 1909 to the Concorde disaster on the outskirts of Paris in 2000. But how does the aeronautics company Aim Aviation (Fliteform) fit in? What is the significance of Vladimir Putin's presidency of Russia? And who is Wm Gray?

Items are linked on the basis of shared dates: 31 December, 25 July, 1 February. This is the classic stuff of conspiracy theory: the idea that dates themselves carry meaning, that there is no such thing as coincidence. Behind the surface of history, behind the apparently random events of past and future, a dark machinery is at work, hidden and always malevolent. Things don't just happen. Someone is pulling the strings and making the puppet dance. Or, more likely, a secret society of such individuals, shadowy but superbly organised and immensely powerful. It's a seductive idea. Dan Brown wouldn't have got far without it.

*

The message in a bottle can be a functional means of communication. In the days before satellite and telegraph, the bottles sometimes contained SOS messages, but this must surely have been a desperate last throw of the dice, given the capricious nature of winds and currents, and the zillion-to-one odds of discovery in time, and by someone with the wherewithal to effect a rescue.

A message famously entrusted to the sea by Christopher Columbus, whose ship had been engulfed in a terrible storm, was designed to let Queen Isabella of Spain know about his discovery of the New World. He feared that if he and his sailors did not make it home the great news would die with them. In fact he survived to tell the tale himself, and the bottle was never found. Perhaps it's still out there somewhere, bobbing unnoticed in the wake of oblivious cruise ships.

A century later, messages in bottles had become an accepted way for the Navy to send ashore vital military information. For some time this early method of ship-to-shore communication was crucially important. There was even an official Uncorker of Ocean Bottles, appointed by Elizabeth I. No other person was allowed by law to open these official secrets, on penalty of death.

There are instances of bottles containing eyewitness accounts which document extreme experiences the world might not otherwise understand. They can even be a final communication, or a farewell to the world. After the sinking of the *Lusitania* it was said that a fisherman had plucked from the water a message in a bottle, describing the final moments before the ship disappeared under the sea. Apparently, the letter read: 'still on deck with a few people. The last boats have left. We are sinking fast. Some men near me are praying with a priest. The end is near. Maybe this note will' – the message cut short, presumably as the writer suddenly ran out of time, stuffed it hurriedly into the bottle and cast it overboard.

I don't know whether this is a true story. No one seems to know the whereabouts of the letter, and it's got the whiff of legend about it, not least because there are so many different versions doing the rounds. In 1931 a bottle encrusted with shellfish is said to have washed up on the German coast, and been found to contain a letter signed by ten people on board, saying that the *Lusitania* was minutes . from sinking.

Apocryphal or not, these stories have been circulating in the public imagination since the very beginning – the *New York Times* published a version on 20 July, just nine weeks after the disaster, under the headline *Lusitania Farewell Ashore in a Bottle*. The brief story tells that the bottle was washed ashore near the village of Koudekerke in Holland, and the message inside read as follows: 'Lusitania. We are torpedoed, one in front and one in back. I take leave of my parents and my girl, who live in London, John Street, 57 East End. He who finds this is begged to give this to them. The boat sinks. Farewell forever. J.H. BURTON.'

The newspaper article goes on to say that although there is no J. H. Burton in the published lists of passengers and crew of the *Lusitania*, these lists are known to be 'not entirely complete'. When I consult the 1911 Census, I find there was indeed a family named Burton living at 57 John Street, St Pancras, though whether this could feasibly have been described as 'East End', even in 1911, is dubious. The family comprised Charles (aged 56), Charles (20), Daisy (25) and Florrie (22). No mention of J.H., although since this is just four years before the *Lusitania* disaster he may well already have been away at sea by then. Alternatively, of course, the whole story may have been based on hearsay or journalistic licence.

Whatever the facts, such stories must have been very powerful at the time. People often wish fervently to hear a message from 'the other side', especially when death has come suddenly or violently or as the result of a great catastrophe. Whether or not the *Lusitania* letters were genuine, they would have appealed to that familiar desire for some kind of contact. They suggest a scenario in which the victim, though close to catastrophe, was composed enough to write a letter; in which pen and paper and bottle were still to hand, and there was time and calm enough to bid a proper farewell to loved ones. Heartbreaking though it must have been, perhaps this narrative was easier to bear than the alternative vision of final moments spent in utter chaos and horror.

⋆

Entrusting your final words to the ocean, in the wild hope that someone might find them and pass them on, is an understandable act of desperation. More generally, though, what makes someone go to the trouble of sealing a letter in a bottle and throwing it in the sea for a random stranger to find?

As the Samaritans know well, sometimes it's easier to talk to a stranger. People in desperate situations will call and discuss their problems anonymously down the phone line, when talking about them face to face with their nearest and dearest feels impossible. Or maybe their nearest and dearest just aren't listening.

Are we as a species susceptible to moments of cosmic loneliness, which make us feel that our actual relationships are not enough and send us searching for another, more serendipitous kind of contact? Is there a sense in which every message in a bottle, however apparently lighthearted or banal, is, in Sting's words, 'an SOS to the world'?

Perhaps even the world is not enough for us these days. A cursory search throws up dozens of websites offering instructions on how to make random contact not with other earthlings but with extra-terrestrials. You needn't even seal that bottle and go down to the sea; it can be done from the comfort of your own desk. One site offers to transmit your message for you to 'one of the scientifically selected locations' by means of 'pulsing photonic emission and radio frequency transmission'. On the not unreasonable question of how likely the message is to get through, you might be encouraged by the assurance that 'our algorithmic calculations show that your chance of making contact using our proprietary technology are far greater than your chance of winning the lottery'. If that's not enough, here's the clincher: 'You don't get a Limited Edition signed certificate suitable for framing when you play the lottery.'

Not everyone thinks we should be looking for aliens. Professor Stephen Hawking admits it's 'perfectly rational' to assume intelligent life exists elsewhere, but he thinks we should do everything we can to avoid contact with it. 'If aliens visit us,' he says, 'the outcome would be much as when Columbus landed in America, which didn't turn out well for the Native Americans . . . we only have to look at ourselves to see how intelligent life might develop into something we wouldn't want to meet.'

In fact, though, the business of sending messages to alien life forms has been attempted, on behalf of all of us, and with the firm stamp of official approval. These messages, often characterised as interstellar messages in bottles, have been blasted into outer space on a number of occasions. The most famous, known as the Golden Record, was placed aboard the Voyager spacecraft in 1977, and is now travelling through empty space. If it's ever found, and whoever finds it can interpret the diagrammatic instructions on how to play it, they will experience a series of images and sounds which were specially selected to describe life on our planet, including recordings of birdsong, thunder, whales, the music of Bach and Chuck Berry, pictures of humans and animals, greetings in fifty-five languages, and special messages from President Jimmy Carter and UN Secretary-General Kurt Waldheim. However, it only got beyond our solar system in 2004, and it will be another forty thousand years before it approaches any other planetary system. All those voices, groping their way through the dark for forty thousand years – there's a lonely thought, if ever there was one.

Most of the messages in bottles, too, are bobbing about somewhere, their words of friendly greeting or desperate appeal for help lost in the intergalactic spaces of the world's oceans. In the days before plastic, bottles were more likely to be smashed to pieces on rocks than found intact on the shore; but whatever the vessel, the odds are against it finding its way. As we know, a large proportion of the world's rubbish ends up floating about in the North Pacific Gyre; there must be thousands of messages in bottles there, each one still stoppered and sealed. Columbus's frantic dispatch to the Queen of Spain might be mixed up with them too. Possibly half a dozen *Lusitania* letters. Hordes of silent *cris de coeur*, swept up into the Great Pacific Garbage Patch. A colossal heap of undelivered mail.

*

I resort to amateur graphology in an attempt to crack the code and understand this message. My research tells me that when someone writes entirely in capital letters it's because they don't want who they truly are to be seen. Capitals are the least revealing of all writing styles, apparently. The classic ransom note is always written in capitals.

I am advised, too, to take the slant and the heaviness of the writing as an indication of how deeply emotions affect this writer. The letters here are scrupulously upright, and very heavy indeed.

I know I'm grasping at straws. But what else have I got to go on?

Perhaps it was never meant to be read at all. Some messages are driven by more private motives. They are, in the words of another song, 'letters I've written, never meaning to send'. In this spirit, a French woman, still grieving hard for the son killed in a road accident twenty-one years ago, wrote an anonymous letter to him, sealed it in a glass bottle and threw it from a cross-Channel ferry. She never really intended it to be found; it was more of a letting go, rather like a scattering of ashes. But her bottle, like a heat-seeking missile, made its way straight onto a beach in Kent, and into the hands of a writer, Karen Liebreich, who happens to be fluent in French. It was, of course, irresistible. Liebreich's book, *The Letter in the Bottle*, was published in France, to intense media attention, and before long the mother surfaced, hurt and furious, feeling 'violated'. How could her private act of grief have ended up being turned into something so public? But that's the essence of the message in a bottle. Once you lean over the side of the ship and let go, you relinquish control. It will find its own reader.

I'm sure my message is different. There's a peculiarly controlled sense of intensity to it. I'm sure it was meant to be read; after all, it found me twice.

But that's just the start. A speculative search on the internet reveals, to my astonishment, that I am not the only person to have found this message. A blogger, Simon Jones, wrote about it in 2008, and several people responded to say that they had found one too. Between them, they recounted three or four versions of the text, all overlapping, variations on a theme. Along with the air crashes, some mention 'The Umbrella Man', 'The Radioactive Man' or the death of Princess Diana. Sometimes the initials DMTNT appear at the foot of the page: *Dead Men Tell No Tales*.

All sorts of things can become the subject of conspiracy theories. According to one poll, 47 per cent of Americans believe that the Apollo moon landings never really happened but were elaborately staged. There is a recurring claim that Paul McCartney died in a road accident in 1966 and was replaced by a lookalike. Air crashes, though, are the quintessential stuff of the conspiracy theory. Korean Airlines Flight 007, shot down in Soviet airspace in 1983; Pan Am flight 103, blown up over Lockerbie in 1988; TWA Flight 800 brought down off the US coast near New York in 1996: these are just a few of the most feverishly disputed disasters. There are allegations of elaborate cover-ups and whitewashes. Cynical attempts to mislead the public. Secret photographs and taped phone calls. Darkness at the heart of government.

No one is immune to the lure of the conspiracy theory, and it would certainly be a kind of madness always to believe the official line, no matter what. But for some people, signs of conspiracy crop up everywhere and even start to take over their lives. An obsessive compulsion to believe, prove, or re-tell a conspiracy theory can be a symptom of a more general paranoia. The thread on Simon Jones's page is rife with speculation about the meaning of the messages, the possible source, the psychology behind it. Some people think it's a hoax, others interpret it as a cry for help. Why would anyone write this stuff? What kind of disturbed person keeps throwing messages into the sea? How can it be normal to send letters to people you don't know? Eventually someone with a sense of irony writes, 'Aren't all blogs just nothing more than messages in a bottle?'

One message in a bottle feels like a gift. Two starts to look like junk mail. Someone is mass-producing these things and spamming them out there, on the principle that in the end he'll get a response. But given that his messages are anonymous, what would that response look like? If he found a fellow believer, how would he ever know? Perhaps it's enough to feel he's alerting the world to what's really going on. And of course he must do so in code, because you never know who might intercept the message.

Sadly, my mystery correspondent is not talking to the whole world. He has a relatively limited audience. A few of his bottles have been washed up in Ireland, Scotland and Wales, but most have been found on the north-west coast of England: Fleetwood, Whitehaven, Formby, Greenodd, Aldingham, Morecambe, Overton, Bayliff, the Lune estuary, Cleveleys, Poulton-le-Fylde. He's certainly not going to trouble any of the current long-distance record-holders, like the bottle thrown overboard by a crew of German scientists in the middle of the South Indian Ocean, which made a 16,000-mile journey in six years. Or the one sent by a schoolgirl in Lancashire, who received a reply from a boy nine thousand miles away in Perth, Western Australia. Incidentally, that particular bottle began its journey in the Lancashire town of Heysham, postcode LA3. Wait a minute, this sounds like the makings of a conspiracy . . .

*

Whoever wrote these messages, he – I'm certain it's a he – has gone to a lot of trouble. They are artefacts, with their own curious beauty. The care with which the paper has been removed from the pad, the text written and annotated, the thread tied and the bottle sealed suggests a labour of love. I'm reminded, as I look at them, of the hobby of putting ships in bottles – the fine work that goes into rigging the masts, attaching the sails with strings and hinges so the masts can lie flat against the deck, then, once the ship is inside the bottle, tweezering them up and into position.

But to go through the ritual of making the same artefact, month after month, for several years – that must take dedication. Especially when it gets chucked in the sea as soon as it's finished. At least the finished ship in a bottle goes on the mantelpiece, to be admired by family and friends.

I don't know how long my two bottles have spent in the sea, but the papers inside are both dated – one 4/08/10 and one 12/11/09. Months, then, rather than years. Once washed up on the beach, they lay amongst the rich debris of the tideline, pecked experiment-ally by gulls, blown along the beach a little by a gust of wind. They were locals, compared with much of what lies around them today: a fruit basket from Thailand; a Korean flask, with its mystery contents still sealed inside; an empty packet of Mirage lime mixed tobacco from India; a French condom wrapper. The whole world comes to this beach. The tideline is an open book in a babble of different languages: an account of what the world desires, and then wishes to be rid of. My beautiful, mad, cryptic messages belong here like everything else.

First-footing

And once, in some swamp-forest, these
Were trees.
Before the first fox thought to run,
These dead black chips were one
Green net to hold the sun.

These are the opening lines of 'Coal Fire' by the American poet and anthologist Louis Untermeyer. It's included in his *Golden Treasury of Poetry*, first published in 1959. Throughout my childhood it was my favourite book, and this page was one I turned to again and again, both for the poem and for the spooky Joan Walsh Anglund illustration of a gnarled and split tree, with roots like claws gripping the rock face, against a wash of strange green light.

Introducing the poem, Untermeyer says it is 'another instance of how poetry can be made out of seemingly unpoetic material'. Coal is such ordinary stuff – dirty, heavy, boringly familiar – and yet at the same time so miraculous and alchemical. It begins life as a tree, which dies and falls into water, becomes peat bog, is covered by heavy sediment and metamorphoses into rock. What we dig out of the ground is stored energy: heat from the sun, packed tight and ready to be released when we set light to it.

A strong tide brings a scattering of coal to the beach. It looks different here. If you pick it up, you find it washed to a shine. Some pieces are as small as pebbles, others like house bricks. I had one block as big as a pulpit bible. It took two of us to carry it home, and it sat on the hearth for a year before we lit a ceremonial fire on New Year's Eve and wedged it, with some difficulty, into the grate.

It was a pale Southern imitation of the Scots tradition of 'first-footing', where the first visitor to the house after midnight – preferably a tall, dark stranger – brings a lump of coal as a gift, signifying warmth for the year to come. Coal is a thing of mixed fortunes, representing both good and bad luck in different traditions and circumstances. Kissing a chimney sweep or touching his hat or the buttons of his coat is considered lucky in many European countries. On the other hand, no child wants to be judged so badly behaved that all they find in their stocking on Christmas morning is a measly piece of coal. It's magical stuff; but it's humble too. We're not sure what to make of it.

A few years ago, I wrote a poem of my own in which I imagined tracing the origins of the coal I was finding washed up on the beach. The poem is a fantasy, in which a man dreams of taking a boat far out to sea and using it as a kind of time machine, diving through the past to search for the drowned forest which was the original source of the coal:

> And there it is: an undisturbed continent of trees.
> The forest floor swells with life.
> The canopy bubbles with birds and monkeys
> adapted to life underwater.
> There is sunlight in the crucible of leaves.

At the time, I had little idea where the beach coal might be coming from. When I asked around, no one else seemed to know either. I was vaguely aware that leftover coal sometimes washes out of disused mines, and I thought that might be the case here, so I had my invented character start out by asking himself, as he walks along the beach, 'What redundant seam wasted this spoil?'

*

A piece of petrified wood found on the beach is like a prototype for coal. It's a type of fossil, in which the wood has turned to stone but still keeps its shape and structure.

I don't know where my piece came from, but remains of pre-historic forest have been emerging from the sea around these parts for hundreds of years. The Proceedings of the Liverpool Geological Society in 1871 contain a paper by the Society's founder, George Highfield Morton, bearing the rather verbose title 'The Progress of Geological Research in Connection with the Geology of the Country around Liverpool'. The article includes a description of 'peaty outcrops' containing tree trunks on the shore, and quotes an eighteenth-century description of the discovery of a 'submarine forest'. (Incidentally, this volume of the Society's proceedings also contains a letter from one Charles Darwin, entitled, somewhat more snappily, 'Tidal Action as a Geological Cause'.)

There are still remnants of submarine forest visible around the mouth of the River Alt at very low tide. There was once extensive marshland here, known locally as 'the moss' or 'the black earth'. In 1615, William Webb, Clerk to the Mayor's Courts of Chester, writing about Dove Point on the Wirral coast, said: 'In these mosses, especially in the black, are fur trees found underground, in some places six feet deep or more, and in others not one foot, which trees are of surprising length and straight, having certain small branches like boughs, etc.'

Or as his contemporary the Reverend Richard James put it, less bureaucratically and more decoratively, in his poem 'Iter Lancastrense':

> in summe places, when ye sea doth bate
> Downe from ye shoare, 'tis wonder to relate
> How many thowsands of theis trees now stande
> Black broken on their rootes, which once drie land
> Did cover, whence turfs Neptune yields to showe
> He did not allways to theis borders flowe.

*

It turns out that the dream scenario in my poem bears some distant relation to the facts. The coal washed up here can be traced back to a submarine seam beneath the Irish Sea. There are seams like this in various places around the British coastline, and many beaches which are studded with coal after a high tide. In some places the supply is so plentiful that collecting it can still be a commercial proposition; 'seacoalers' have worked along parts of the north-east coast of England for generations. Horse and cart have given way to the ex-army Land Rover, but it's still back-breakingly hard work, shovelling and loading the coal by hand.

So extensive are some of the undersea seams that one energy company has been surveying to investigate the potential for coal exploitation at five offshore sites, including Cromer, Sunderland and Swansea Bay. The company estimates the total quantity of coal in the five areas at one billion tons. Since this coal is not accessible for mining, it would be 'gasified' by drilling boreholes and injecting a mixture of water and oxygen.

The Irish Sea is not on the list, so for now the sporadic supply of coal to this beach will continue, and I'll go on picking it up occasionally for the fire. It's a contemplative pursuit. Each piece has its own rough-hewn character, its own heft and shape. They're like huge uncut gems, chiselled from the ground, raw and glinting. The sea has smoothed their sharp edges a little. They smell of salt, and they're clean as if they were newly made.

Rough Lords

There's plenty of death here on the beach. Dead sheep and dead seals are fairly common sights, and there was a dead minke whale a year or so back, with a deep wound to its abdomen where it had been in collision with a boat. And that's just the big stuff. Zoom in close, and you can see that the foreshore is littered with casualties, especially after a storm: crabs, starfish, cockles, clams, jellies, worms and urchins, in their last silent throes, or already carrion.

Still, a dead bird is a peculiarly shocking sight. People have always envied birds their flight: the freedom to soar in open space, above the rush and clutter of the ground, to look down on it all. Ever since Icarus, we have been using our human ingenuity to try and achieve the same thing – sticking on feathers, attaching wings to bicycles, making flying machines with balloons and jet engines, leaping out of them with parachutes. Birds, on the other hand, find it all so easy and natural. They seem almost godlike in their mastery. And yet here it is, that free spirit of the air, brought down to earth and undone, just as mortal as any of us.

The young herring gull at my feet is very much undone. It's as if someone has unpicked the stitches which held it together, and spread it out like a broken book. The neck twisted to one side, the bill open. A spoonful of viscera. A wing lifting in the breeze, like a page printed with paragraphs – pitch, slate, smoke, salt – and only the wind skim-reading them.

*

One of the most frequently recurring and enduring images of happiness, especially romantic happiness, is of a beach, often at sunset, and often, it must be said, backed by palm trees. Nowhere in Britain is more than seventy-five miles from the coast, but our own native beaches – often rainy and windswept – have a more ambiguous significance in the popular imagination, partly because of our climate and also, no doubt, because they are so close to home. It can be harder to see the attraction.

But for many of us – those of my generation and older – our own beaches too have a life in the world of memory and fantasy: they epitomise the childhood holiday. I grew up in the most land-locked part of England, and we would see the sea only once a year, on the annual family holiday. We would compete to be the first to spot it from the back windows of our old Ford Cortina; that scrap of blue, glinting on the horizon a moment and then lost behind hills or trees, represented the zenith of excitement and anticipation. In that instant, there was *everything* to look forward to.

These holidays mostly took place in North Wales, surrounded by the mountains and wild open countryside my parents loved. We also had a fortnight of mixed favours in Scarborough in the early seventies, when my brother and I remember playing on the beach in thick fog. Spells of bad weather were inevitable, and would not be allowed to prevent our beach fun. Come rain or shine, we'd be there: picnicking, playing French cricket, building forts for the tide to conquer. If you were cold, you simply put on a jumper over your swimsuit and ran about more. That bracing attitude has stayed with me to this day: I'm not immune to the obvious attractions of a beach in summer, but I have my happiest adventures there in winter, when more fastidious souls are curled up by the fire at home.

But what about those lucky people who grew up by the seaside? What does it mean to them? In this town, attitudes are mixed. Some people are fiercely proud of their coast, with its Sites of Special Scientific Interest and Special Protection Areas, its National Trust and Statutory Nature Reserves. But there are plenty more who never visit. It's often blow-ins like me who get all worked up about it. In the nineties a major new development was built, right on the seafront – one of those anonymous complexes with shops and restaurants and a cinema – but to my surprise it was built with its back to the sea. It's called Ocean Plaza, but day-trippers walking along the promenade see only the blind concrete backs of the buildings; there isn't a single café or balcony there with an ocean view.

My walk has brought me to an area known locally as the Green Beach, or Smith's Slack. The sand here is being progressively colonised by vegetation, a process which was observed and recorded in the 1980s. A decision was made to allow nature to take its course here, and the result is a remarkable biodiversity which grows richer every year. The natural process accelerated when car parking at the foot of the sand dunes was prohibited. By 2002, this area was in transition between saltmarsh and sand dune. Research found that the greening of Smith's Slack provided a more stable foreshore, which offered greater protection against storms and flooding.

It hasn't been a universally popular development amongst the townspeople, however. The green stuff is still perceived by some as weeds, which are burgeoning out of control, spoiling the beach, making it scruffy and overgrown. What's to stop it spreading and spreading until it overwhelms the clean, smooth sands which every seaside town feels are its number one asset? Even now, twenty-five years on, there are occasional angry demands for the council to get off its backside and tidy up, just as a suburban neighbour might try to insist that the bloke down the street cuts back his hedge and mows his lawn.

But leaving this area to change and develop in its own way has been a great success for the coastal species that thrive here. Walking on this section of the coast, you can see no fewer than seven distinct types of habitat: strandline, embryo dunes, saltmarsh, fixed dunes, freshwater marsh, dune slack and alder scrub. Between them, they provide a rich variety of conditions which support a range of rare and threatened plants, birds and animals.

Here at the top of the beach, embryo dunes are forming. On a windy day like this, you can see how it all starts: loose, dry sand is blowing onshore and catching on clumps of marram grass and other strandline material. If you lie down so that you're looking at them sideways on rather than from above, you can see the miniature dune beginning its life. Trapped here, the sand accumulates into tiny peaks. Gradually these peaks will be colonised by salt-tolerant plants, which further slow the wind speed and catch more drifting sand. The succession continues, with the dunes growing so that they are no longer covered even by the highest tide; at this point they become known as 'mobile dunes'. Sand is blown onto them and they are patchy with vegetation. Once they are completely colonised, they become 'fixed dunes'. By walking up the beach here, you can see this whole process of succession. It's like watching something being born and growing up, right before your eyes.

Once over the embryo dunes and the ribbon of saltmarsh, I'm

out on the wildest, most remote part of the beach. You can walk here for a couple of hours without seeing another person; this is far from the vehicular entrance, backed by high dunes and much less accessible to the casual walker. It's a zone which really belongs to the seabirds. I'm a visitor in their territory now, and I think of Richard Mabey's warning about wild places: 'If we go into them it should be as a privilege, and on the same terms as the creatures that live there, unarmed and on foot'.

The wildness and variety of bird life embody the things I love about this coast. My ornithological credentials are nothing special – I observe birds with curiosity when I'm walking, and usually forget to read up about them when I get home. But even with my slipshod ways, I've managed to identify a reasonable range of seabirds over the years, including sandpiper, oystercatcher, dunlin, snow bunting, redshank, sanderling, cormorant, lapwing, kitti-wake and bar-tailed godwit, as well as common, black-headed, Mediterranean, herring, glaucous and lesser black-backed gulls.

There's a richly layered aural texture here. The bass note is a steady background roar which seems to come from deep inside the sea's machinery. Then there's the percussive sound of each wave breaking, like a word being spoken, and again, and again, but every time with a slightly different accent. If you listen really hard, you can hear the longer, quieter phrase which is the wave climbing, rolling, falling back. And over it all, the sharp sentences of the gulls' cries: five or six syllables, varied and repeated.

⋆

On the way through the dunes today I saw – and then heard – my first skylarks of the year. It's less usual to encounter them that way around. The classic way to experience skylarks is to lie in the grass

206

and let the sound seep into you along with the sunshine, warming your bones and your weary soul. It seems paradoxical, but these exalted birds make their nests on the ground, in tufts of grass. Today I frightened one from its nest into the air, where it began that characteristic hovering rise and rise, higher and higher until it was hard to see as I screwed up my eyes against the sun – it was just a speck against the limewash sky. As it climbed, it paid out that unmistakable sustained, rolling song – 'His rash-fresh re-winded new-skeinèd score', as Gerard Manley Hopkins describes it – which is part of the soundtrack of summer, a delicious promise on this February day.

The skylark sounds quintessentially English, a quality Vaughan Williams reached for in his impressionistic orchestral piece The Lark Ascending. This Classic FM favourite must be among the best-known of all attempts to render bird-music into formal human-music. Then there's Beethoven's 'Pastoral' Symphony, which includes snatches of song from the nightingale (represented by the flute), quail (oboe) and cuckoo (clarinet); the cuckoo is recognised easily enough, though I've always felt a bit doubtful about the quail.

The birdsong in the 'Pastoral' Symphony may have been a novelty item for Beethoven. The French composer Olivier Messiaen, on the other hand, had a lifelong obsession with birdsong. He claimed that birds were the greatest of all musicians, and considered himself as much an ornithologist as a composer. He transcribed the songs of birds all over the world, from the Jura in his native country to Japan and the Middle East. His piece Le Réveil des oiseaux is entirely built from birdsong: it's a dawn chorus for orchestra. Birdsong interested Messiaen on more than one level: not only did it evoke a landscape, a scene, a season and a mood, but it provided structure

207

and pattern through its complex repetitions, improvisations and variations on a theme.

When you lie in bed with the window open in the early hours, listening to the real-life dawn chorus, it takes time to tune in. For the first minute or two it sounds chaotic, a jumble of competing sounds. Then, gradually, different lines of melody come into focus. It becomes possible to pick out individual voices, to hear variations in tone, pitch and texture, and to notice repetition, ornamentation, call and response. As David Rothenberg writes in his book *Why Birds Sing*, 'no one who listens would call a bird's sound random'.

But what does it all mean? That's the big question, the one which continues to fascinate and frustrate human listeners. 'The language of birds is very ancient,' says the great eighteenth-century naturalist Gilbert White, 'and, like other ancient modes of speech, very elliptical: little is said, but much is meant and understood.' We may teach ourselves to recognise particular calls and the functions they fulfil – warning, mating, territorial, etc. – but it's clear that this is just the start. All those complexities and variations defy simple reductive explanations. Rothenberg, after repeated sessions jamming along on the clarinet with birds in the US National Aviary in Pittsburgh, reaches the conclusion that not all birdsong is straightforwardly functional. 'Why do birds sing?' he asks again; and suggests one possible answer, persuasive in its simplicity: 'For the same reasons we sing – because we can. Because we love to inhabit the pure realms of sound.'

While composers have been fascinated by the challenge of representing birdsong in human music, others have tried to put it into words. From childhood I've always loved reading the transcriptions in bird books. I'm not talking about approximations to rhythm using borrowed words and phrases – the yellowhammer's alleged request for *a little bit of bread and no cheese* – but real attempts to give a sense of the sound. It's a project which calls for concise and meticulous writing. Opening the *Collins Bird Guide*, I find, for example, that the blackbird has a song 'well known for its melodic, mellow tone, a clear and loud fluting (almost in the major key) at slow tempo and on wide, often sliding scale, with soft twitter appended'. The song of the mistle thrush is said to be similar, but with a 'more desolate and slightly harder tone'.

Many of our garden birds are unquestionably singers. Seabirds also make extensive use of their voices, but it would be stretching a point to describe the sounds they make as 'song'. Their cries ring out forcefully against the background noise of the wind and the waves, powerful and unequivocal. Herring gulls can be recognised by what the British naturalist Mark Cocker describes as their 'gales of manic laughter'. It's not tuneful; it's a sound built to carry across distance, a sound that fills the space, that announces ownership.

According to the *Collins Guide*, the black-headed gull makes a 'strident, downslurred, single or repeated "krreearr", with many variations, and short, sharp "kek" or "kekekek"'. Rather like a human infant, the young little tern begs for food 'with a light, ringing "plee" or "plee-we"'. The Mediterranean gull, on the other hand, is described in terms uncannily familiar to any parent of teenagers: 'Call is distinctive, rising-then-falling note, like enthusiastic, nasal "yeah!", but with slightly whining tone'.

*

Nowadays, some bird music is heard less often. The dawn chorus is recreated morning after morning, but the texture is changing.

In *Waterlog*, Roger Deakin refers back to his childhood collection of I-SPY books, which listed things to spot, and awarded points according to rarity value. 'It is interesting to compare how rare or common things were perceived to be in the 1950s, compared to our present-day perceptions. In my I-SPY Birds, I find that the linnet and the song thrush score a mere twenty points, level pegging with the starling and the house sparrow.' I certainly can't think when I last saw (or heard) a linnet. Along with the song thrush, it has been on the Red List of endangered species for some time. Now even the once ubiquitous sparrow has joined them, and is described by the RSPB as 'struggling to survive in the UK'. This is a heavy blow for a bird so utterly familiar, one which regularly features in the top ten of favourite British birds.

Herring gulls are never going to make it into the top ten, but I'm certain they couldn't care less. The Liverpool poet Matt Simpson wrote of them as

> scavengers, rough lords
> of rubbish dumps who strut on garbage,
> dip beaks in waters rank with sewage
> where the city stains its tides
> blood-brown, mobbing wakes
> of ferries, persecuting cargo boats
> for galley slops

– and in the past fifty years they have extended their domain well beyond ports and seaside towns. Their numbers are actually declining, but people see more of them because they're being forced inland by loss of natural habitat and dwindling fish stocks. They're highly adaptable birds, colonising sewage outflows, landfill sites, town centres, and wherever else the pickings are rich enough. Attempts to control them are mostly futile – they attack bird-scaring devices, and perch on special anti-perching spikes. The *Daily Mail* recently described them as 'noisy, filthy and violent', citing cases in which 'lunatic' gulls stalk and dive-bomb people, and rip the lead flashing from their roofs.

These are essentially urban problems, but it's certainly not uncommon for a beach visitor to be targeted by a hungry gull mounting a blitzkrieg for his fish and chips. Everything is fair game, and their opportunistic feeding style is on display every day here at the beach. Herring gulls will eat fish, crabs, invertebrates, small mammals and other birds. Neither are they particularly fussy whether any of these are alive or dead; they are happy to kill and to scavenge, and are not averse to cannibalism. According to *Birds Britannica*, they are known to ingest 'rope, bonfire charcoal, match-boxes, greaseproof paper and the rubber seals around car sun roofs'. But as Pablo Neruda observes in his 'Ode to the Sea Gull', there's a kind of alchemy at work; no matter how much trash the gull consumes,

> it is transformed
> into clean wing,
> white geometry,
> the ecstatic line of flight.

Nothing is wasted here: not the decaying remains of seaweed, not dead crabs and starfish, not the fish-heads discarded by anglers. Seagulls are natural scavengers and play their part in clearing up beach debris. But inevitably their turn comes too. So what about the dead gull itself? What will happen to these unstitched wings, still sporting their juvenile colours? What will become of the leathery feet, the gaping beak, the clotted blood, the eyes?

They'll be eaten, of course, like everything else. First on the scene will be more gulls – just try keeping them away. Then crabs and other crustaceans. The odd passing dog, maybe. Flies and bacteria will finish the job.

Or perhaps, before the corpse is completely consumed, a high tide will sweep in and claim it. Once in the sea it will be food for a different range of scavengers, and what particles are left will fall as 'marine snow' to the darker depths, where they will be picked up on the sticky tentacles of brittle stars and sea anemones.

*

I leave the corpse behind and head for the sea. I pass close to a gang of live gulls, solid and stoical. They hunch their shoulders and watch the sea.

Today, here on this loneliest and least frequented part of the shore, there are hundreds of oystercatchers, their black-and-white plumage and long, tangerine-orange bills bright and distinctive against the backdrop of beach and sky. They are feasting on mussels, chiselling them open with their long, blunt bills. Another flock comes streaming in to land: a host of small barrel-shaped bodies propelled on quick wingbeats. They make a very distinctive flight call, a shrill, clear cry as cold as the winter waves.

Out here there are flocks of sandpipers too, huddled very close together. Sandpipers keep up a perpetually anxious, quavering sound – the poet Elizabeth Bishop describes them as 'in state of controlled panic'. They and I engage in a game of brinkmanship, in which I get as close as I can without startling them into flight. They watch me intently, bubbling collectively in readiness. Then – on some imperceptible signal – they rise as if they are not many but one, perfectly co-ordinated, and as they turn back against the glassy blue of the sky, and then back again, they seem to be something other than matter: at first I think they must be made of light, and then I think no, they're sound. The quicksilver-white of them one way, the stormy black of them the other.

Sights such as this dazzling, switching shoal of birds can momentarily alter our preconceptions about separateness and collectivity. It makes me think, suddenly, of Fritjof Capra's book *The Turning Point* (1982), in which he suggests that the planet is made up not of individuals but of systems. Ultimately, he describes the whole planet as a single overarching system, or ecology. It's a development of James Lovelock's *Gaia* hypothesis, which describes 'a complex entity involving the Earth's biosphere, atmosphere, oceans, and soil; the totality constituting a feedback or cybernetic system which seeks an optimal physical and chemical environment for life on this planet'. Capra goes further still, extending this holistic way of thinking beyond the strictly scientific to include politics, psychology, medicine and economics.

The way in which we describe the natural world and our place within it has changed radically since *The Turning Point* was published. At that time, environmentalism was still a minority interest, despite the fact that Rachel Carson's iconic book *Silent Spring*, and the campaign group Friends of the Earth, were both in their third decade by then. It has taken time for the ideas of those early thinkers and activists to move from somewhere out on the hippy fringe to the very centre of political, social and cultural life.

Thought does not necessarily equal action; we go on hurtling towards ecological meltdown. But the *ideas* have taken such a firm hold that it's difficult to think of any context in which they have no presence. Our very ways of thinking, speaking, reading and watching are conditioned by them, and it has become virtually impossible for us to talk to one another about 'nature' without at the very least a tinnitus of anxiety. We use a neurotic vocabulary of loss and the fear of loss; we utter only elegy or omen.

It often seems that simple enjoyment of the natural world is a luxury we can no longer afford. Aren't we all nostalgic for a time when it was possible to listen to birdsong without hearing gaps and silences; or to walk along the seafront in a storm without thinking of climate change?

But this is like hankering after the simplicities of childhood, which we know in our hearts are irretrievably lost.

White Horses

It's a squally day of dangerous skies and sudden daggering light. A navy-blue horizon where the Welsh hills should be. On the ground, a smirr of glittering wormcasts, each packed with fragments of white shell.

Everything else is jittery with movement. A few feathers skitter across the surface, and tattered rags of grey froth slide northwards, like a platoon of ghosts, translucent and shape-shifting.

It's good weather for kiters. You need strength, stamina and courage to manage the big kites. I was on the beach in gale-force winds back in April, and I watched a kiter struggling to stay on his feet; harnessed to the kite, he shared its fortunes, and the wind had that scrap of engineered nylon between its teeth and was not letting go. In the end he took the last resort, flinging himself down onto his back on the sand. Still the kite did not give up. It was lifting him by the harness and dropping him hard on the sand, lifting and dropping, as if he were nothing more substantial than a piece of litter.

This beach was the site of an attempt to break the land speed record in 1925, and nowadays landsailers use traction kites to race wheeled buggies. Those who prefer to get their adrenalin rush on water go for kitesurfing. You stand on a board with foot straps or bindings and use the power of a large controllable kite to propel you and the board across the water at terrifying speed. It's an extreme sport, requiring expert handling of the kite as well as skilful use of a surfboard. Your body is the only connection between the two, and you have to control them both at the same time: piloting the kite on the sky and steering the board on the water. Wind powers you along, and if you get really good at it you can make 'big air jumps' of fifty feet or higher.

Out there, the great curved kites are strokes of colour against the stony sky, and the sea is crested with 'white horses'. Here the tide is romping in, driving whipped froth onto the shore.

Foam on the beach gets people worried, but it's usually nothing to do with pollution. When algal blooms break down far out at sea, organic material is released into the water, and this very fine protein suspension acts as a foaming agent. It is churned up in the breaking waves, coats the air bubbles and makes a stiff froth, like the head of foam in a beer glass. Only if it has an oily texture or a bad smell is it likely to be caused by something more sinister.

The 'foam of perilous seas' is a sign of the ocean's dynamic, irrepressible character. To John Keats, it has an otherworldly quality, evoked by the nightingale's song, seen from 'magic casements' in 'faery lands forlorn'. To John Masefield, 'the flung spray and the blown spume' speak of adventure and freedom. In Charles Kingsley's 'The Sands of Dee' the wind is 'wild and dank with foam' on the day Mary drowns, and the foam itself becomes a monster: 'cruel', 'crawling' and 'hungry'.

In some parts of the world, it's certainly a force to be reckoned with. Where there are rocky shores and pounding surf, really colossal amounts of foam occasionally gather, piling up several metres deep on the shore and making beaches inaccessible. Beaches in New South Wales have been inundated several times in recent years, and in high winds the froth is blown inland and covers houses, gardens and roads.

Here, under a menacing sky, it's on a much smaller scale, creeping and spectral. Bits of ectoplasm break loose and roll away on their own. Others catch on shells and cling on, ashen and shivering. But it's mercurial stuff, taking its mood from light and shadow. The sun cuts through for a moment, and suddenly it sparkles festively, like scattered meringue.

Time Travel

Nine o'clock on a January morning. I crunch my way through sand dunes hardened and sheened with frost, then slither over a sheet of ice, which is the winter beach. Under the ice, pale bubbles swell and skitter away from my tread. The tideline is an ice-line: a sparkling white ribbon of frozen froth, curling away into the distance ahead and behind. I stop and watch oystercatchers pecking at a frozen pool. I visit a remote shipwreck, its timbers rimed with frost. I walk on, hard and fast, trying to work up some warmth.

After a couple of miles, I'm just thinking of heading inland and going home for coffee, when I spot a large expanse of broken ground, darker than the surrounding sand. The tide is very low, and this broken area is well down the beach. As I get closer, I'm shouting out loud with astonishment, which is fine because there is no one else in sight: I have this whole spectacular discovery to myself. It's an area perhaps fifty metres by thirty: complex and layered, startlingly different from the smoothness of the beach behind me. There is a series of miniature cliffs, islands and peninsulas, all made of dark-brown stuff like clay, moated around and between by channels and pools of icy water which the sea has left behind.

This strange and incongruous sight is full of special significance. It's a piece of the distant past, resurfacing right in front of my eyes. The muddy outcrops I'm looking at are so ancient that they offer the chance to visit a world lost thousands of years ago, before 'history' began. But it's a fleeting opportunity, before the next tide comes in and erases it for ever.

This is not an accidental find, but it is miraculous. It hasn't been cast up randomly on the tide; nevertheless it lay hidden and undiscovered until the 1950s, when people began to notice animal hoofprints appearing from time to time in sun-hardened outcrops of mud along the foreshore. Even then, their significance was not understood; until 1989, when a local man called Gordon Roberts began walking his daughter's dog and observing them for himself. Most of us, looking at those prints, wouldn't have had a clue what they were. We'd hardly have noticed. They wouldn't have seemed out of the ordinary, and we'd have walked on without a second thought. There's no shortage of footprints on the beach, after all; think of all the people, dogs, horses and birds that traverse the shore, day in, day out, leaving their prints as evidence until the next tide comes in.

But Gordon recognised that these were special. He'd had a lifelong interest in archaeology, and he understood straight away that they were likely to be very old. He began to make phone calls to universities and museums. He set about making a photographic record and carrying out his own research.

To begin with, it was suggested that the prints might belong to domesticated cattle kept by Iron Age people. There is a theory that a sub-tribe of the Brigantes called the Setantii may have settled this coast, all the way from the Mersey north to Borrow Beck in Cumbria. Cattle footprints would have been an exciting piece of evidence in support of the theory. A discovery of well-preserved animal prints two thousand years old would certainly have been something to write home about.

However, as time went on, these early ideas about their provenance looked less convincing. Gordon returned again and again to the site, using his experience of archaeological practice to record what he found in a systematic way, making notes on the exact locations, taking measurements and photographing the evidence. He observed variations, depending on weather and tidal conditions: one day there would be nothing visible, and the next he would find new exposures, more and different prints. Birds and animals of various kinds. And in amongst them, *Homo sapiens*.

Meanwhile, plaster-casting of the hoofprints allowed them to be taken away and studied in detail, and they just weren't consistent with the other evidence we have to tell us what Iron Age life was like. Iron Age people were farmers rather than hunters; the animals they kept were cattle, sheep, oxen and pigs. Yet the hoofprint analysis was identifying quite a different range of species, including roe deer, red deer and wild boar. This was no farmyard.

*

The discovery of the footprints began to draw attention and attract scientific investigation. Samples of the silt in which they were found were taken and subjected to both carbon dating and a newer technique called Optically Stimulated Luminescence, which can determine how long ago sediment was last exposed to daylight by measuring radiation in the minerals it contains. These processes gave dates from the late Mesolithic (7,000 BC) to the early to late Neolithic (5,000 BC). This means that the earliest prints were laid down around the time that Neolithic culture was beginning to emerge in Northern Europe. Neolithic innovations included farming and herding, but in places like this, unsuitable for agriculture, people went on living by hunting, gathering and fishing until much later. The wild boar and deer that left their marks here would have provided food for these communities, and the human footprints are those of the hunter-gatherers who relied on them.

The discovery brought about a seismic shift in the way historians thought about this place. Until now, there had been no hard archaeological evidence that this stretch of coast had been occupied by people of any kind until Norwegian settlers arrived in the tenth century AD. The prints, and the painstaking work to date them, changed the received wisdom. They were proof that prehistoric people had been here, taking advantage of the rich pickings: gathering seafood such as shrimps and clams, and pursuing deer and wild boar. Indeed, so rich were the pickings that the hunter-gatherer lifestyle persisted here for many hundreds of years after people in other areas of Britain had begun to settle and farm the land.

This evidence is available to us only because of a lucky combination of accidents in climate and coastal change. The prints have survived here because they are preserved in layers of sediment which until recently were hidden under the surface of the sand. The sediment is what remains of a reed-fringed lagoon and mudflats which once occupied this area, where wild animals roamed and people hunted them for food.

The shape of the coastline has always gone through periods of change, as the sea advances and retreats. During one of these periods, the sea level fell and the coastline moved out westwards, covering over and sealing in the mudflats. For three and a half thousand years the sediments, and the secrets they contained, lay buried beneath.

In the early twentieth century, another period of change got under way, and erosion of the dune coast began. Tidal conditions had changed, perhaps due in part to the widening of the shipping channel in the Mersey estuary. For a hundred years now the coastline has been retreating as the sea pushes it back. The prehistoric mud now lies under the beach instead. It's very shallowly buried, sometimes only about six inches beneath the surface of the upper foreshore, and is easily exposed.

In periods of relative calm, a system of ridges and runnels forms; deep runnels cut down through the sand, exposing sections of the ancient sediment underneath. The sediment is very dark in colour, almost black, and completely different in texture to the sand around it. When the wave energy is low, you can see it in places, swirling about at the edge of the sea and streaking the sand with black dust. Sometimes people think it's oil. As you run barefoot towards the edge of the sea, you can find yourself slithering in a patch of black mud which oozes up from under the sand. It's prehistory, stirred up after all this time, carried to the surface and coating your feet.

Once you know what you're looking for, you can spot these dark, broken outcrops from a distance. Close up, there's a stepped effect, with a number of distinct layers or strata exposed, one under the next. Here on a good day, you'll find footprints. The chances are you'll be the first person to see them.

What I've stumbled across today is a remnant of the ancient mud lagoon, and it's much more extensive than the fragments I've seen before. It's deep, and made up of many strata. Each might represent a hundred years, or five hundred; I can't say; but they're quite distinct to the eye. If you break off a piece, you can peel them apart; they're flexible as rubber. And within each stratum there are countless micro-strata, each one representing, perhaps, a twice-daily silt-laden tidal incursion over a period of some two and a half thousand years.

The remnant I'm looking at is on one of four main sites in an area known as Mad Wharf. Human and animal prints emerge at these four sites from time to time, and according to a largely unpredictable timetable of weather and longshore currents. The heyday was the mid to late 1990s; as the sea continues to move inland it has covered much of the former lagoon area. Fewer and fewer prints are being exposed, and no one knows how far inland they reach, and how many more remain to be discovered.

I'm slipping and sliding around on the surface, just as people did thousands of years ago, only with the advantage of walking boots. Although actually I'm not so sure they *are* an advantage – they keep getting stuck, and I have to pull them out with a squelching sound like a spoon from a trifle. It's close to freezing, but I decide to take them off so that I can get around more easily and feel the mud between my toes like my predecessors did.

*

The image of footprints in the sand is a very popular and enduring one. They can represent an idealised image of love, trailing behind the couple who are strolling hand in hand on the perpetual white-sand beach of our dreams. It's an image which simultaneously captures the eternal nature of human experience (we make marks as we go) and its transience (the tide will come in and wash the marks away).

Perhaps the principal significance of footprints in the sand is as a sign that we need never walk alone. But what if we *want* to walk alone? Surely there's something attractive about the promise of untouched sand, or the safety of solitude? What speaks of a comforting presence to one person may spell a terrifying threat to another, as it does for Robinson Crusoe, whose discovery of a single footprint in the sand was anything but reassuring:

It happened one day, about noon, going towards my boat, I was exceedingly surprised with the print of a man's naked foot on the shore, which was very plain to be seen in the sand. I stood like one thunder-struck, or as if I had seen an apparition. I listened, I looked round me, but I could hear nothing, nor see anything; I went up to a rising ground, to look farther; I went up the shore and down the shore, but it was all one; I could see no other impression but that one, I went to it again to see if there were any more, and to observe if it might not be my fancy; but there was no room for that, for there was exactly the print of a foot – toes, heel, and every part of a foot . . . Nor is it possible to describe how many various shapes my affrighted imagination represented things to me in, how many wild ideas were found every moment in my fancy, and what strange, unaccountable whimsies came into my thoughts by the way.

Crusoe has to face the terrifying knowledge that he is not alone on his desert island. His first thought is that the devil must have visited. Eventually he comes to his senses and realises it's just another man. But instead of rejoicing at this thought, he's driven into still greater agonies of dread. He may have been desperate with loneliness in the past, but by now he feels safe. Suddenly he realises that company isn't all it's cracked up to be.

*

Here, alone on my island of prehistoric mud, I too find my first human print. It is perfectly defined, deeply indented where the foot has sunk several centimetres into the soft surface, and lined with tiny shells. And now my eye is accustomed and I can see dozens more, developing before my eyes like photographs in a darkroom. Sometimes there's just a single, vaguely foot-like shape I identify by squinting and looking from different angles and hoping for the best. Others are as clear as if they were made yesterday. And here's something much more exciting: a distinct trail of imprints, emerging from underneath one of the layers and leading out towards the sea. Trails like this are the really valuable evidence; they have an 'integrity' you can't assume of a single print. It's too easy to mistake the recent marks left by a dog or a child for their ancient counterparts. Some of the prints are filled with water; one still has a skin of ice which wrinkles when I touch it with a fingertip. All are as big as my own foot, or bigger, so I suppose they were made by adults. Gordon has found children's footprints too, including what he has described as 'a turmoil of tiny, sun-hardened footprints' which evoked for him the image of a group of children 'stomping around in the cool mud – quite literally "mud-larking" – while their elders foraged for shell food nearby'.

Alongside the slow cycle of coastal accretion and erosion there are seasonal patterns, and daily changes in weather conditions, which keep things unpredictable. You can find an exciting new set of prints and mark them with plastic sticks, intending to return, only to be thwarted by windy weather, which blows sand over the site, or heavy rain, which damages the prints beyond recognition. Or they may be degraded or polluted by vehicles and horses crossing the mud.

These frustrations are an inescapable fact of life in what Gordon calls 'ephemeral archaeology'. Our email correspondence has been full of provisional arrangements and last-minute cancellations. He wrote excitedly one day to tell me about his discovery of possible wolf prints, but no sooner were the prints spotted than we had a day of summer storms and they were obliterated for ever.

Gordon is guru of the prints, and guide to the prehistoric lagoon. When he first took me there, on a cold, bright morning a few days before Christmas, the early signs were not encouraging. An onshore wind had blown sand into the runnels, and the beach surface stretched away flat, smooth and uniform. Still, we walked on, and after ten minutes we came to a small patch of exposed sediment containing some red deer hoofprints. They were surprisingly large; red deer were huge compared with today's species. Their hooves could be as long as fifteen centimetres. There were quite a few of their prints exposed that day, and amongst them – quite suddenly – a human footprint. The outline was a bit blurred, but the shape was unmistakable.

I felt a surge of excitement. There was no one else about, and the two of us were almost certainly the first people to see this footprint since it was made, thousands of years ago. 'This is what thrills me most,' said Gordon, digging into his rucksack for his camera and a wooden ruler: 'tickling the toes of our ancestors.' But the camera batteries were out of charge, and by the time he'd replaced them a wave had swept in, inundated the prints and very nearly washed away the ruler. Nothing would survive of them. So we were the last, as well as the first, to see this particular piece of evidence. For Gordon, it was frustrating but quite normal. Without measurements and photographs, we could only head back to the National Trust HQ, warm up with mugs of tea and talk about what we'd seen, and what we hadn't.

On the way, I asked him about his discovery and the difference it had made to his life. He's something of a local celebrity these days, and his work is increasingly well known nationally and internationally, especially since it was featured on the BBC series *Coast*. The television producers may have put photogenic young presenters in front of the cameras, but the story they told came straight from Gordon. At an age when others might have settled into quiet retirement, he's involved in cutting-edge research, writing about his findings, speaking at conferences and leading guided walks on the shore. It's a subject which still makes his heart beat faster, and when I ask him what his most exciting moment has been he answers without a moment's hesitation, 'The antler'. A young woman on one of his walks spotted what she thought was a bone sticking up out of the sand. He began to brush away the sand, and the bone got bigger and bigger, and as more of its distinctive surface texture was exposed he became sure it wasn't a bone after all. The tide was coming in fast, so they agreed to return the following day and investigate further. With great care they dug out the object, which was quickly identified as a red deer antler and later dated at five thousand years old.

What's wonderful about these stories is the improbable nature of the discoveries, occurring as they do against all the odds. For every footprint revealed to us there are others which will forever lie concealed; and the footsteps which left prints behind are just a fraction of those which were made in the first place. For prints to persist, certain conditions were required. First, the original consistency of the sediment had to be soft and muddy enough for an indentation to be made and to remain after the foot was removed and its owner had walked on. Next, the print would be covered by very fine, light, windblown sand. The weather must have been warm and fine, allowing the imprinted sediment to dry out sufficiently before the next tide came in and covered it. Finally, the tide itself needed to be gentle and with very little wave action, so that the surface was not disturbed but delicately sealed with silt and clay.

*

Another animal which left its mark here was the aurochs. In Britain this mighty creature, a species of wild cattle, had been hunted to extinction by the end of the Bronze Age. It was a real monster, standing two metres high at the shoulder and weighing a thousand kilograms. Julius Caesar wrote about the aurochs in his *Gallic War*, describing it as 'a little below the elephant in size, and of the appearance, colour, and shape of a bull'.

Because it has been extinct for hundreds of years, the aurochs prints found on the beach are highly prized for the insights they offer into the animal's lifestyle, its physical characteristics, and even the way it walked as it grazed on the dune grasses and wallowed in the mud. There is some skeletal evidence elsewhere: the skull of the last surviving aurochs is in a museum in Stockholm, and remains have been excavated in Peterborough. But much of the information about this beast comes from a strange and unexpected source: the Palaeolithic cave paintings at Lascaux in the South of France. Indeed, the detailed depiction of a hoofprint on one of these paintings played a key role in helping confirm the identification of the aurochs tracks here on this beach.

There is something marvellous about the potential of these paintings, made sixteen thousand years ago, to help in scientific discovery today. We have no written evidence to help us understand the lives of people in that remote past, but we do have some pictures. The fact that archaeologists are willing to extrapolate from them, to use them to build and test hypotheses about the lives of pre-historic people and the creatures which shared their world, speaks eloquently to the quality of prehistoric art. It may have been made without much natural light, with sticks of charcoal and bits of ochre and haematite; nevertheless, it survives and tells detailed stories about a part of our past which would otherwise be unreadable.

Recapitulation theory tells us that there are parallels between the way our distant ancestors did things and the way young children do them. Each of us develops from a fishlike embryo, mirroring the way our species developed from fishlike ancestors. As children, we go from crawling to walking upright, we learn to use tools, and so on. *Ontogeny recapitulates phylogeny.* This is sometimes misunderstood, so that in the popular imagination there is an idea that a simple progression can be traced from the relative incompetence of the past to the genius of our own time. So persistent is this idea that we are amazed, over and over again, by the skill and sophistication of cave paintings. These are not rough sketches, analogous to the child's crayon scribble; but detailed renderings, anatomically correct and specific enough to distinguish between closely related species, and between male and female, young and old. This is down to the skill of the artist, but it also speaks of the intimate knowledge prehistoric people had of the animals they hunted, butchered and ate. Australian Aboriginal art even includes a distinctive style known as X-ray, which depicts the outline of the animal (often a fish or another marine creature such as a turtle) containing diagrams of the bones and internal organs.

In acknowledging the expertise of these early artists, we get to the heart of the debate about the meaning and purpose of the drawings they made. Drawing may be a complex creative activity, but it's also functional. When we draw, we concentrate on the things that matter to us or that we need to know. A child drawing a cat may do a pretty good job of indicating the shape of its ears and tail, and perhaps even draw in a whiskery smile to express the cat's apparently happy disposition. The resulting drawing records what the child knows – and needs to know – about cats, or this cat in particular. The prehistoric hunter, on the other hand, knew a bison inside out – and needed to, since his survival depended on it. Interestingly, drawings of humans on cave walls are relatively rare, and where they do occur they are often rough schematic figures, oddly out of place amongst the complex and realistic representations of animals.

Everyone is fascinated by cave paintings, but nobody knows why they were made. The evidence is that most of these caves were never lived in, so it's unlikely that the painting of the walls was purely decorative in purpose, although some people do argue that it may have been a form of graffiti. There's plenty of informed speculation about hunting magic, shamanism and even astronomy: parts of Lascaux have been painted with what appear to be startlingly accurate star maps. Then there's a mass of other theories – less scientifically respectable, but even more fervently expressed – that they contain coded messages about the location of the Holy Grail, for instance, or document visits to Earth by aliens from other planets. The truth is frustratingly inaccessible; in spite of our common humanity, we are so distant in time from the people who made these images that we're left with few interpretative tools other than guesswork and extrapolation.

Like the footprints I'm looking at now, the cave paintings at Lascaux were sealed away and forgotten for thousands of years, unvisited and unknown. The discovery of the cave, like the discovery of the footprints or the buried antler, was a wonderful accident. The story has been told, retold and mythologised over the years, but it's so extraordinary, so touching and so true to our collective fantasies of childhood that it really needs no extra ornamentation. Even the simplest version reads like a psychedelic version of an Enid Blyton story. On 12 September 1940, four local teenagers and their dog, Robot, were out treasure-hunting when Robot fell down a hole. (There's always a pet dog in stories like these, running off and leading its owners unwittingly into the adventure; though whether Robot survived, history does not relate.) In the search for him, the boys fell into a deep cavern which they'd never known was there, though they knew the area intimately. It was too dark to continue the search, so they clambered out, and came back the next day with a rope and a torch. It must have been an unforgettable moment when they lowered themselves into the hole in the earth, shining the torch into the darkness, and saw for the first time those cave walls, vividly decorated with magnificently preserved paintings of animals, humans and strange abstract symbols.

I've never been to Lascaux. It was closed to the public in 1963, after the body heat and breath of over a million visitors had begun to cause serious damage to the paintings. Photographs show the range of the images: horses, bison, buffalo, a bear, a bird. And here's the aurochs itself: a solid, bull-like creature with powerful shoulders and long, curved horns.

I did once visit another of the painted caves of southern France, Font-de-Gaume, while on holiday with friends. There is no aurochs pictured here, but my visit gave me a glimpse of the anatomical detail and accuracy of Palaeolithic art. You enter the cave through a narrow gap under a cliff, and make your way down a twisting tunnel. We went in from the incandescent 40 degrees of an August heatwave, and the cold and gloom were like the slam of a church door. For the first few minutes I couldn't see much at all. Then, very gradually, the first few images 'developed' before my eyes, and after a while I could see that there were hundreds of them, all around me. There were bison, mammoth, horse, reindeer, goat, wolf, bear and rhinoceros. The highlight was a frieze of five bison, which uses shading to suggest three-dimensionality, an astonishingly sophisticated technique for the Ice Age artist. As I stared, depth and perspective seemed to rise to meet me. It was a little like gazing at those 'magic eye' stereograms, where you refocus your eyes to see the three-dimensional image emerge from the page.

Art historians have drawn attention to similarities between the draughtsmanship on display in these caves and techniques associated with Modernism. Legend has it that Picasso, the greatest and grandest of Modernist painters, emerged from Lascaux after a visit in 1940, and pronounced to the waiting audience: 'We have invented *nothing*.'

*

Perhaps we have the wrong idea about beaches. Golden sands are pretty; but mud is full of treasure. Ancient pollen grains preserved there tell us what plants were growing in the remote past. Traces of charcoal suggest campfires and settlement. And then there are the footprints.

They have yielded sufficiently detailed information both to give a general idea of life in Neolithic times and to build profiles of the individuals whose feet made them. Gender is relatively easy to distinguish, because men's and women's feet are surprisingly different: not only do men's tend to be longer and broader than women's, but a woman's foot has a higher arch, a shallower first toe, a smaller ball-of-foot circumference and a smaller instep. With expertise, it's possible to reach conclusions about the way someone moved: to estimate stature, relative stride, velocity, stride time, cadence and speed of walking. Evidence of an unusual gait can be analysed and used to suggest that a woman was at an advanced stage of pregnancy, or that another was suffering from bursitis or a foot deformity characteristic of diabetes or muscular dystrophy.

It's an incredible wealth of information from a series of very old dents in the sand, and it collapses time, giving us a tantalising glimpse of the actual people who walked here, where I'm walking today. Like the cave paintings at Lascaux or Font-de-Gaume, they confront us with the oddly surprising realisation that these were men and women like us. G. K. Chesterton, writing about Lascaux soon after it was discovered, argued that it overturned previous ideas about the nature of prehistoric man:

> What was found in the cave was not the club, the horrible gory club notched with the number of women it had knocked on the head . . . If any gentleman wants to knock a woman about, he can surely be a cad without taking away the character of the cave-man, about whom we know next to nothing except what we can gather from a few harmless and pleasing pictures on a wall.

The 'cave-man', then, is not the stereotypical club-wielding brute, but the artist, the thinker, the man. I say 'man', but recent work has revealed that women were cave painters too: some of the handprint signatures found on cave walls have been measured and studied and found to be consistent with female hands.

Those cave handprints, like these beach footprints, are graphic physical messages from the past. Sherds of pottery, fragments of cloth, even drawings are all part of the material culture: telling, but secondary. Where a human being has placed a hand or a foot, and made a mark, there's a short cut of recognition, a spark leaping across millennia. What connects us is that we share the same kind of body.

So here I am, alone on the shore on this freezing January day, looking back at a time before recorded history began. As I stand staring at the shapes of the feet of long-dead, unknowable people, I'm reminded of that strange, unsettling image from the Hubble Space Telescope: the Ultra Deep Field image, which peers back thirteen billion years and captures a litter of oddly shaped galaxies that existed shortly after the Big Bang and are now vanished. The contexts and timescales are wildly different, but the effect is similar: a spine-tingling realisation that now is not all; that this time in which we find ourselves, and which means everything to us, is a random and fleeting moment of negligible consequence. There's a refocusing of the lens, and a new recognition of my own individual insignificance.

Like Crusoe, I've found proof that I'm not alone. It's not just space we share, but time too. When I unlace my boots and step barefoot onto the freezing mud, I experience a tangible sense of connection with the past. There were other lives lived out in this place, and the intertidal zone is the place where their mysteries are kept.

Seven thousand winters have passed since the earliest of these footprints were laid down, preserved and buried. What was the wider context at the time? There was no wheel, and no writing yet. Stonehenge was still a couple of thousand years off. But in Mesopotamia, wheat and flax were being farmed; fine glazed pottery with stylised figures of animals and birds was being made; and the hilltop temple complex at Göbekli Tepe was already an ancient monument at four thousand years old. The lens refocuses again, and for a second I see that four thousand years is not some abstract concept, but simple and actual and not so unimaginably long; it's just that our own lifespan is pitifully short.

And now at last they have come to the surface again: the marks my analogues made, as they slithered across the mudflats gathering shellfish and hunting wild animals for food, and of their children as they ran about and played in the mud. Miraculously, they are here, almost on my doorstep, and I'm lucky enough to be in the right place, at exactly the right moment to witness them before they vanish for ever. I'm standing here, as I stood in the cave at Font-de-Gaume, right in the middle of my blip of a life, with the deep past all around me. I can trace a footprint with my fingers, put my own bare foot right inside it. It's the nearest I can get to time travel.

Postscript

My old walking boots – the ones with the eyelets rusted brown by seawater – are now in a cupboard in my new home in London. They still have a little dry sand in the treads, a remembrance of my year on the beach.

I have other souvenirs too: a china cup, a piece of petrified wood, a toy duck, a sea squirt shaped like a human ear. Even the Neolithic footprints, the most fugitive of my finds, are represented here by a slab of prehistoric mud which has dried to the colour of old stone.

These pieces of the beach have made the move with me, and have a place in my new life now. Lined up on my mantelpiece, they form a peculiar little museum, which I tour from time to time, picking them up and breathing their faint but unmistakeable scent of the sea. They are loosely connected, this assortment of objects, by the time they spent jostling together in the water or tangled in the strandline, and by the way their stories touch and overlap.

My beach has its own character, but there are similar discoveries to be made on beaches everywhere, from an unidentified creature wriggling on the wet sand to a miscellany of lost and abandoned human possessions. Things arrive unannounced, then disappear again under the waves; buried history comes to the surface; traces of the past are exposed and erased. Beaches can be intensely cosmopolitan places; and they can be like time machines.

Each tide brings in another cargo of mysteries; there's always something, once you get used to looking. But the real thrill is in the chance nature of these encounters. Like so many of our happiest meetings, they are coincidental. You happen to be walking along the right part of the shore, just as something is delivered there by the tide. The two of you are on separate journeys. You come from one direction, it comes from another, and your paths intersect.

Acknowledgements

I am deeply indebted to all those who have offered advice, expert information and much-needed encouragement during the writing of this book; and especially to those who have walked with me and shared the excitement of discovery.

Particular thanks to Stephanie Anderson, David Barrie, Andrew Brockbank, Mandy Coe, Diane Davis, Barbara Dutton, Martyn Griffiths, Matthew Hollis, Simon Hollis, Nigel Pantling, Laurence Rankin, Gordon Roberts, Robin Robertson, Vera Di Campli San Vito, Mark Sargant, Martha Sprackland, Thomas Sprackland, Richard Villa and Ric Williams.

I acknowledge with gratitude a Roger Deakin award from the Society of Authors, which helped make this book possible.

Credits

Lines from 'The Sun-Fish', in *The Sun-Fish* by Eiléan Ní Chuilleanáin (2009), by kind permission of the author and The Gallery Press. Lines from 'Crazy About Her Shrimp', in *The Voice at 3:00 a.m.: Selected Late & New Poems* by Charles Simic, copyright © 2003 by Charles Simic. Reprinted by permission of Houghton Mifflin Harcourt Publishing Company. All rights reserved. Lines from 'Sea Mouse' and 'Beach Glass', in *The Collected Poems of Amy Clampitt* by Amy Clampitt, copyright © 1997 by the Estate of Amy Clampitt. Used by permission of Alfred A. Knopf, a division of Random House, Inc. Lines from *The Dead Seal near McClure's Beach* by Robert Bly courtesy of Sceptre. Lines from 'Rough Hollow Pearls in the Seaweed Night', in *The Curse of the Killer Hedge* by David Bateman (IRON Press, 1996), by kind permission of the author. Lines from 'A Wreck', in *Corpus* by Michael Symmons Roberts (Jonathan Cape), reprinted by permission of The Random House Group Ltd. Lines from 'Snow', in *The Collected Poems of Louis MacNeice* by Louis MacNeice (Faber & Faber), by kind permission of David Higham Associates. Lines from 'The Park Drunk', in *Swithering* by Robin Robertson (Picador, an imprint of Pan Macmillan, London), copyright © Robin Robertson, 2006. Lyrics from 'Nights in White Satin' by Moody Blues, written by Justin Hayward, copyright © Tyler Music Ltd. Lines from 'Coal Fire', in *Golden Treasury of Poetry* by Louis Untermeyer, by kind permission of the Estate of Louis Untermeyer,